KANT'S INAUGURAL DISSERTATION
AND
EARLY WRITINGS ON SPACE

KANT'S INAUGURAL DISSERTATION AND EARLY WRITINGS ON SPACE

Translated by

JOHN HANDYSIDE

M.A. (Edin.), B.A. (Oxon.); Late Lecturer in Philosophy in the University of Liverpool

HYPERION PRESS, INC.

Westport, Connecticut

Published in 1929 by Open Court Pub. Co., Chicago

Hyperion reprint edition 1979 1988
Library of Congress Catalog Number 78-59026
ISBN 0-88355-699-5
Printed in the United States of America

Library of Congress Cataloging in Publication Data
Kant, Immanuel, 1724-1804.
Kant's inaugural dissertation and early writings on space.
Reprint of the 1929 ed. published by Open Court Pub. Co., Chicago.
1. Reality. 2. Space and time. I. Handyside, John, 1883-1916. II. Title. III. Title: Inaugural dissertation and early writings on space.
B2794.D42E5 1979 115'.4 78-59026
ISBN 0-88355-699-5

PREFACE

The only original writings left by Mr John Handyside—*The Historical Method and Other Essays*—have been published by[1] Professor Pringle-Pattison and Professor Alexander Mair, with an account of the author's career and heroic death. Among his papers have also been found a translation of Kant's *Inaugural Dissertation* and of the treatise on *The First Ground of the Distinction of Regions in Space*, together with a few pages of what was evidently designed to be an introduction to Kant's early writings, or at least to those which deal with space and the methods employed in mathematical science. These papers are here printed, and should prove a valuable addition to philosophical literature in the English language. Kant's *Dissertation* has hitherto been strangely neglected, largely, no doubt, owing to the very difficult, and often obscure, Latin in which it is written. Yet it is a very helpful propaedeutic to the *Critique of Pure Reason*, and quite indispensable to any genuine understanding of Kant's mental history and of the influences which, already in 1770, were preparing the way for his appreciation, some two or three years later, of Hume's sceptical questionings.

A translation of the *Dissertation*, by Dr William J. Eckoff, has indeed appeared in English, but in a not very accessible form, in the publications (1894) of Columbia University, New York. Though having considerable merit, it will not easily bear comparison with Mr Handyside's translation. Mr Handyside was, however,

[1] Liverpool, The University Press, 1921.

acquainted with it, and doubtless found it occasionally helpful.

I have added a translation of those sections in Kant's first published writing, *Thoughts on the True Estimation of Living Forces*, which embody his earliest views with regard to space or which are otherwise relevant as bearing upon problems dealt with in the *Dissertation*. A few notes which have been appended, are distinguished from Kant's own, by enclosure in square brackets.

It has not seemed desirable to attempt to complete Mr. Handyside's Introduction. This, as it stands, serves as an admirable introductory note to the translations.

In preparing these pages for the press, I have been indebted for assistance to Professor Pringle-Pattison, Professor Alexander Mair, Professor A. E. Heath, and Mr Henry Barker.

NORMAN KEMP SMITH

UNIVERSITY OF EDINBURGH
JUNE, 1927

CONTENTS

INTRODUCTION

The recurrent point of all Kant's early speculations first appears in the *Thoughts on the True Estimation of Living Forces*. In this treatise, as frequently in his pre-Critical writings, the interest is largely of a physical order. Kant is concerned with substances, physical in their nature. We can ask, in regard to any physical substance, whether or not it is connected with any other physical substances. "Either substance is in a connection and relation with others outside itself or it is not. Since every self-sufficient being contains within itself the complete source of all its determinations, it is not necessary for its existence that it should stand in relation to other things. Substances can therefore exist, and yet have no outer relation to others, nor stand in any actual connection with them."[1]

In order to grasp the possible metaphysical bearings of this problem, we must pass far beyond Kant's actual horizon at the moment. We must bear in mind his later problem, how we can come to knowledge of an universal connection between an antecedent and a consequent, that is, to knowledge of a law of cause and effect. This problem is not here stated in its full universality. We may perhaps, as an approximation, say that what first occupied Kant's attention was the problem of *transeunt* action. Assuming that we can think a plurality of substances in their independence and as substances (we may note the largeness of this assumption), have we any

[1] *Op. cit.*, § 7. Cf. below, p. 8.

ground in thought for asserting that these substances, in their independence, have (as Kant puts it later) any *commercium*, any interchange of activities? Further, if we start with the substances as intelligibles, and regard space, as Kant did at the time, as a subsidiary phenomenon, dependent on the intelligible relations of these substances, it is no more essential to them that they should have spatial position than that they should stand in intelligible connection.

The contrast between this and Kant's later view will at once be noted. In the Critical period, what we may call the *de facto* universality of space and time, conceived as forms of sensibility, becomes the medium for the application of the forms of intelligible law, and for the proof of the *de jure* universality of those forms. The order of dependence is thus reversed; and accordingly, in reference to the things (i.e., phenomena) dealt with in the later philosophy, there is no room for repetition of the question so disturbing in the earlier philosophy: how independent things come to be connected. For the things are no longer the independently conceived substances of intellect, but the conjointly perceived objects of sense. Their connection, so far as it is a problem, is part of the same problem as their substantiality.

For the present, however, Kant holds to his principles, and even carries them to further consequences. It is metaphysically true, he proceeds, that more than one world may exist. Millions of independent worlds may have been created; and it remains undetermined whether they actually exist or not. Further, "it is easily proved that there would be no space and no extension, if substances had no force whereby they can act outside themselves. For without a force of this kind there is no connection, without this connection no order, and without this order no space."[1] Again, space being a consequence of the connection of substances, the special character of space as three-dimensional must, it would seem, find its explanation in the special character or law of the connections of substances. But what that is Kant declares himself unable to say (though now, as later, he has suggestions to make); and in consequence he declares, logically enough, that space of other dimensions and other properties is possible, and that "a science of all these possible kinds of space would undoubtedly be the highest enterprise which a finite understanding could undertake in the field of geometry."[2]

All these are subjects which recur in Kant's succeeding works, and have interest on that account. What follows, and constitutes the main part of the work, namely, the discussion whether the measure of force is to be taken (with Descartes) as the velocity, or (with Leibniz) as the square of the velocity, is of less philosophical value. In Part III, however, the views taken in regard to the validity of mathematics have considerable interest owing to the respects in which they differ from any of Kant's views at other periods. Thus, when,[3] in consequence of results reached in the above discussion, the question is raised how a law found false in mathematics can hold good in nature, Kant replies that the axioms of mathematics are artificial presuppositions, and that the conclusions come to are valid only within the limits which

[1] *Op. cit.*, § 9. Cf. below, p. 10.

[2] *Op. cit.*, § 10. Cf. below, p. 12.

[3] *Op. cit.*, § 114. Cf. below, p. 13f.

these presuppositions prescribe. "Mathematics determines its concept of body by means of axioms which it insists upon presupposing in its treatment of body; and these axioms are of such a character that they do not permit, indeed they exclude, certain properties which are yet necessarily met with in body as it exists in nature."[1] This is interesting as formulating a view of mathematical knowledge which Kant did not himself continue to hold;[2] although it is one which has found favour with more recent thinkers on the subject.

[1] *Loc. cit.*

[2] [Cf. the views which Kant developed in 1764 in his *Enquiry into the Clearness of the Principles of Natural Theology and Morals*, and which he repeats in the *Critique of Pure Reason* (A, 712–38; B, 740–66).]

THOUGHTS ON THE TRUE ESTIMATION OF LIVING FORCES

(SELECTED PASSAGES)

[1747]

THOUGHTS ON THE TRUE ESTIMATION OF LIVING FORCES[1]

(Selected Passages)

PART I

§ 1. Every body has a force essential to it.

My purpose, which is that of establishing, with certainty and finality, the doctrine of living forces, will, I thing, be furthered, if I begin by defining certain metaphysical concepts bearing on force in bodies in general.

We say that a body which is in motion has a force; for to overcome hindrances, to bend springs, and to displace masses is, as all the world agrees, to act. This force, if we look no further than the senses teach, may be regarded as something which is communicated to a body entirely from without, and in which the body does not participate when in a state of rest. Prior to Leibniz all philosophers were of this opinion, with the sole exception of Aristotle, whose obscure entelechy was supposed to contain the

[1] [This treatise, which appeared in 1747, is Kant's first published writing. Its full title is as follows: *Thoughts on the true estimation of living forces, and criticism of the proofs propounded by Herr von Leibniz and other mechanists in their treatment of this controversial subject, together with some introductory remarks bearing upon force in bodies in general.* Kant employs the phrase *lebendige Kraft* as the German equivalent of *vis viva*. The nature of the controversy is outlined by Latta, *Leibniz* (1898), pp. 86–93. To the references there given should be added Leibniz's three contributions to the *Acta Eruditorum*. The controversy was opened by his first contribution in 1686 (*Acta*, pp. 161–3): *Brevis Demonstratio Erroris memorabilis Cartesii. . . .* He replied to objections in a second contribution in 1691. It was in his third article in 1695, entitled *Specimen Dynamicum, pro admirandis Naturae legibus circa Corporum vires et*

secret of the actions of bodies. Though the Schoolmen invariably followed Aristotle, none of them have understood this mysterious teaching; and perhaps it is not so conceived as to be intelligible to anyone. Leibniz, to whom human reason owes so great a debt, has been the first to teach that in body there inheres a force which is essential to it, and which indeed belongs to it prior to its extension. *Est aliquid praeter extensionem, imo extensione prius:* these are his actual words.[1]

§ 2. This force in bodies Leibniz entitled active force.

Leibniz gave to this force the general title of active force.[2] His successors would have been well advised simply to have followed him in his metaphysical doctrines; but the attempt has been made to define this force more definitely. The force in body, it is said, is a *moving* force, since we never observe it to generate anything except motions. If it exerts pressure, it is striving towards motion; but the force is only then exercised when the motion

---

mutuas actiones delegendis, et ad suas causas revocandis, that he first distinguished between dead and living force: "Hinc patet duplicem esse *Nisum*, nempe elementarem seu infinite parvum, quem et *solicitationem* appello, et formatum continuatione seu repetitione Nisuum elementarium, id est impetum ipsum. Quamquam non ideo velim haec Entia Mathematica reapse sic reperiri in natura, sed tantum ad accuratas aestimationes abstractione animi faciendas prodesse. Hinc *Vis* quoque duplex: alia elementaris, quam et *mortuam* appello, quia in ea nondum existit motus, sed tantum solicitatio ad motum, qualis est globi in tubo, aut lapidis in funda, etiam dum adhuc vinculo tenetur; alia vero vis ordinaria est, cum motu actuali conjuncta, quam voco *vivam*." (Acta, pp. 148, 149. Cf. Kant's *Werke*, Berlin ed., Bd. I, pp. 522–24).]

[1] ["In rebus corporeis esse aliquid praeter extensionem, imo extensione prius, alibi admonuimus." This passage occurs in the *Acta Eruditorum* (1695), Leibniz, "Specimen Dynamicum," p. 145.]

[2] [wirkende Kraft.]

is actual. In opposition to this view, I assert that in professing to secure an answer to the question of the cause of motion, by thus ascribing to body an essential moving force (*vis motrix*), we practically resort to the same artifice as the Schoolmen, who in their inquiry into the grounds of heat and cold took refuge in a *vis calorifica aut frigifaciens*.

§ 3. This essential force may justly be entitled *vis activa*.

We do not speak correctly if we treat motion as a kind of action, and so ascribe to it a force synonymous with it. A body to which an infinitely small opposition is made, and one which therefore hardly acts at all, has motion in an especial degree. The motion is only the external phenomenon of the state of the body; for the body is not here acting, and yet is striving to act. If, however, through [meeting] an object, it loses its motion suddenly—that is, in the moment in which it is brought to rest—it is then active. We should not, therefore, take our title for the force of a substance from that which is not an action; and still less should we say of the bodies which act while they are at rest (e.g., a sphere which through its weight presses upon the table on which it lies) that they strive to move. For since in moving they would not be active, we should have to maintain that in so far as a body is active it strives to fall into the state in which it does not act. Accordingly *vis activa* is a much more appropriate title for the force of a body than *vis motrix*.

. . . .

§ 5. The difficulties regarding the action of body upon soul which arise from the view that body has no other force than *vis motrix*.

As we do not clearly apprehend what a body does when it acts in the state of rest, our minds always pass in thought to the motion which would follow if the resistance were removed. For the purpose of securing an external characteristic of what takes place in the body and cannot be seen, this resorting to motion would suffice. But usually the motion is regarded as that which the force does when it gains a free outlet, and as being its sole consequence. This is but a slight departure from the true way of thinking, and easily rectified; but we must not, on that account, regard it as an error of small importance. For while the consequences do not indeed show themselves in mechanics and natural philosophy, in metaphysics it is difficult, with any such teaching, to conceive how matter can be capable, in any genuinely active manner (i.e., through physical influence), of generating representations in the soul of man. What, it is claimed, can matter do beyond causing motions? All its force can at most result merely in displacing the soul from its position in space. How is it possible that the force, which can only give rise to motions, should generate representations and ideas? The latter being things of so entirely different an order from motions, it is not conceivable that they should have their origin in a force of that description.

§ 6. The difficulty which similarly arises regarding the action of soul upon body, and how through the introduction of a *vis activa* it can be removed.

We meet with a similar difficulty when the question is raised, how the soul is capable of setting matter in motion. Both this and the above difficulties vanish, and considerable light is cast upon the [nature of] physical influence, when the force in matter is viewed not in terms of motion but in terms of those effects in other substances which we are not in a position to define more precisely. For the question whether the soul can cause motions, that is, whether it has a moving force, now takes the altered form, whether its essential force can be determined to an outwardly directed action, that is, whether it is capable of acting on other beings outside itself, and so of producing changes in them. To answer this question in a completely decisive manner we have only to bear in mind that the soul, as having position in space, must be able to act outside itself. The concept of that which we entitle position, as we find upon analysing it, itself refers us to the mutual actions of substances. A certain acute writer[1] has been prevented from rendering complete the triumph of physical influence over pre-established harmony solely by this slight confusion of concepts; and yet we easily deliver ourselves from it, immediately the attention is directed to it.

How matter should be able to determine the soul to certain representations is easily conceived when we entitle the force of bodies in general active force.

Equally easy is it to deal with the other type of paradoxical assertions: how it is possible that matter, viewed as being able to cause motions only, should impress certain representations and images upon the soul. Matter, when set in motion, acts upon everything that is spatially connected with it, and therefore also upon the soul; that is, it changes the inner state of the soul in so far as that state stands related to what is outside [the soul]. Now the whole inner state of the soul is nothing save the comprehending together of all its representations and concepts;

[1] [The writer in question has not been identified.]

and in so far as this inner state is related to the external it is entitled *status repraesentativus universi.* Thus matter, by means of the force which it has in its motion,[1] changes that state of the soul whereby the soul represents the world to itself. In this way we understand how matter can impress representations upon the soul.

§ 7. Things can actually exist and yet not be present anywhere in the world.

Digressions are difficult to avoid in a subject so extensive as this; and I must again return to what I have been saying with regard to the force in bodies. Since all connection and relation of substances existing outside one another are derived from the reciprocal actions which their forces exercise upon one another, we may take note of what truths can be deduced from this concept of force. Either substance is in a connection and relation with others outside itself or it is not. Since every self-sufficient being contains within itself the complete source of all its determinations, it is not necessary for its existence that it stand in relation to other things. Substances can therefore exist, and yet have no outer relation to others, nor stand in any actual connection with them. Now since there can be no position without external connections, locations and relations, it is quite possible that a thing actually exists and yet is nowhere present in the world. This paradoxical statement, although it is a consequence, and indeed a very obvious consequence, of the most familiar truths, has not hitherto, so far as I know, been noted by anyone. And other propositions, which are not less remarkable, and which capture the understanding so to speak against its own will, follow from the same source.

[1] [Cf. above, § 3.]

§ 8. **That more than one world (in the strict metaphysical meaning of the term) can exist, is true.**

Since we cannot say that anything is a part of a whole if it stands in no connection with other parts (in that case no distinction could be drawn between an actual and an imaginary union), and since the world is a being whose constituents are actually interconnected, a substance which is connected with nothing in the world does not in any way belong to the world, though in thought it may be something. That is to say, it is not a part of the world. If there are many such beings which stand in no connection with any thing in the world, but have a relation to one another, a quite special whole thence arises: they constitute an altogether separate world. Philosophers are in error, therefore, when they teach in their Schools that more than one single world, in the metaphysical sense of the term, cannot exist. It is actually possible that God has created many millions of worlds, taking the term world in its full metaphysical meaning. Consequently, it remains undecided whether or not they also actually exist. The error into which philosophers have thus fallen undoubtedly arises from their not having paid sufficient attention to the explanation of [what is meant by] the world. By definition, only that which stands in an actual connection with the other things which are in the world can be reckoned as belonging to the world.[1] They forget this limitation when they apply their theorem to all existing things in general.

§ 9. **If the substances had no force whereby they can act outside themselves, there would be no extension, and consequently no space.**

[1] *Mundus est rerum omnium contingentium simultanearum et successivarum inter se connexarum series.*

It is easily proved that there would be no space and no extension, if substances had no force whereby they can act outside themselves. For without a force of this kind there is no connection, without this connection no order, and without this order no space. But it is somewhat difficult to comprehend how, from the law in accordance with which this force in substances acts outside them, the plurality of the dimensions of space follows.

The ground of the threefold dimension of space is still unknown.

Since I detect a circular reasoning in the proof which Herr von Leibniz, somewhere in his *Theodicée*,[1] derives from the number of lines which can be drawn at right angles to one another from a point, I have thought of proving the threefold dimension of extension from what is to be observed in the powers of numbers. Their first three powers are quite simple, and cannot be reduced to any others; whereas the fourth power, as the square of the square, is nothing but a repetition of the second power. But however satisfactorily this property of numbers may appear to account for the threefold dimension of space, it fails to stand the test of application. For in anything representable through the imagination in spatial terms, the fourth power is an impossibility.[2] In geometry no square can be multiplied by itself nor any cube by its root. The necessity of the threefold dimension does not, therefore, rest upon the fact that in assuming additional dimensions we should be doing nothing else but repeat the others (as occurs in the powers of numbers), but rather upon a certain other necessity which I am not as yet in a position to explain.

[1] [*Essais de Theodicée*, § 351.] [2] [Unding.]

§ 10. It is probable that the threefold dimension of space is due to the law according to which the forces in substances act upon one another.

Since everything which is to be found among the qualities of a thing must be capable of being derived from that which contains in itself the most complete ground of the thing itself, the qualities of the extension, and consequently their threefold dimension, will be grounded in the qualities of the force which the substances possess in respect of the things with which they are connected. The force, whereby a substance acts in union with others, cannot be thought apart from a determinate law which reveals itself in the mode of its action. Since the character of these laws according to which the substances act upon one another must also determine the character of the union and composition of a multiplicity of them, the law according to which a whole collection of substances (that is, a space) is measured, in other words, the dimension of the extension, will likewise be due to the laws according to which the substances by means of their essential forces seek to unite themselves.

The threefold dimension seems to arise from the fact that substances in the existing world so act upon one another that the strength of the action holds inversely as the square of the distances.

In consequence of this I hold that the substances in the existing world, of which we are a part, have essential forces of such a kind that in union with one another they extend the sphere of their actions according to the inverse square of their distances; secondly, that owing to this law the whole which thence arises has the property of threefold dimension; thirdly, that this law is arbitrary, and that God could have chosen another, for instance the inverse threefold relation; and lastly, that from a different law an extension with other properties and dimensions would have arisen. A science of all these possible kinds of space would undoubtedly be the highest enterprise which a finite understanding could undertake in the field of geometry. The impossibility, which we observe in ourselves, of representing a space of more than three dimensions seems to me to be due to the fact that our soul receives impressions from without according to the law of the inverse square of the distances, and because its nature is so constituted that not only is it thus affected but that in this same manner it likewise acts outside itself.

§ 11. The condition under which it is probable that there exist many worlds.

If it is possible that there are extensions with other dimensions, it is also very probable that God has somewhere brought them into being; for His works have all the magnitude and manifoldness of which they are capable. Spaces of this kind, however, cannot stand in connection with those of a quite different constitution. Accordingly such spaces would not belong to our world, but must form separate worlds. Although in what precedes I have shown that it is possible that a number of worlds (in the metaphysical sense of the term) may exist together, in the considerations now before us we have, as it seems to me, the sole condition under which it would also be probable that a plurality of worlds actually exist. For if only the one kind of space, that which allows of three dimensions, is possible, the other worlds, which I locate outside that in which we exist, could (so far as regards space) be connected with our own, the spaces being all of one kind. We should therefore have to ask why God has separated our world from the others, seeing that through their connection He would have given to his work a greater perfection. For the greater the amount of connection, the greater are the harmony and unity in the world; whereas, on the other hand, gaps and discontinuities violate the laws of order and perfection. It is not, therefore, probable that a plurality of worlds exist (although it is in itself possible), unless it be that those many kinds of space, of which I have just spoken, are likewise possible.

These thoughts may serve as an outline of an inquiry which I have in prospect. But I cannot deny that I expound them just as they have come into my mind, without that certainty which a more prolonged examination of them would have given. I am therefore ready to reject them, immediately a riper judgment reveals to me their defects.

. . . .

PART III

§ 114. How that law which has been found false in mathematics can hold good in nature.

We have shown in detail that the estimation of forces according to the square [of the velocity] turns out to be false in the mathematical field, and that no other measure of force but the Cartesian can be recognized. Nevertheless, at various points in the previous chapter, we have given the reader occasion for hope that the estimation of force according to the square [of the velocity] can be introduced into nature; and it is now time to fulfil our promise. This undertaking will surprise most of my readers; for it would seem to imply that mathematics is not reliable, and that it is possible to appeal against its pronouncements. But that is not how matters actually stand. If mathematics extended its law over all bodies in general, natural bodies would be included thereunder, and it would be vain to look for any exception. Mathematics, however, determines its concept of body by means of axioms which it insists upon presupposing in its treatment of body; and these axioms are of such a character that they do not permit, indeed they exclude, certain properties which are yet necessarily met with in body as it exists in nature. Accordingly body as mathematically conceived is a thing quite distinct from body as it exists in nature; and statements can be true of the former which cannot be extended to the latter.

§ 115. Distinction between mathematical body and body as it exists in nature, and the laws which apply to each.

Let us now consider the nature of this quality which, though met with in body as it exists in nature, is not permitted in body mathematically treated, and which therefore causes the former to be of a kind altogether different from the latter. Mathematics does not allow in the body with which it deals any force which is not completely produced by that which is the external cause of its motion. That is to say, it does not allow any force in body save in so far as it is caused in it from without. This force can therefore always be found complete and in precisely the same measure[1] in the causes whereby the body is moved. This is a fundamental law of mechanics, and being presupposed, allows no other measure of force than the Cartesian. But, as we shall demonstrate immediately, body

[1] [Genau und in eben demselben Masse.]

as it exists in nature has a quite different constitution. It has in itself a power of itself increasing the force which is awakened in it through the cause of its motion. Consequently it can have degrees of force not due to the outer cause of its motion. And since these can be greater than this cause, they cannot be measured by the Cartesian standard; they are to be estimated in a different manner.

. . . .

ON
THE FIRST GROUND OF THE DISTINCTION OF REGIONS IN SPACE

[1768]

ON THE FIRST GROUND OF THE DISTINCTION OF REGIONS IN SPACE[1]

The celebrated Leibniz enriched the sciences by many actual contributions; but he also entertained numerous greater projects, the execution of which the world has in vain awaited from him. I shall not here pronounce an opinion why this was so; whether his essays seemed to him still too incomplete—a scrupulousness peculiar to men of real worth, and one which has ever and again deprived learning of precious fragments—or whether it was with him as with the great chemists, of whom Boerhaave conjectures that they frequently announced performances as if they had them in their power, while actually they were only giving credence to their skill, believing that the execution could not miscarry, if they once resolved to undertake it. At least it seems probable that a certain body of mathematical teaching, which Leibniz in anticipation entitled *Analysis Situs*,[2] never existed save in intention. Many writers, including Buffon (in dealing with the involution of Nature in germs), have lamented the loss which we thereby suffer.

[1] [This treatise was published in 1768. For a statement and discussion of Kant's changing views with regard to the argument here developed, see Vaihinger, *Commentar zu Kant's Kritik der reinen Vernunft*, Bd. II, pp. 518 ff.; Kemp Smith, *Commentary to Kant's Critique of Pure Reason*, pp. 161 ff.]

[2] [The programme of this new science Leibniz outlines in a fragment published by Gerhardt, *Leibnizens mathematische Schriften*, Bd. V, pp. 178–83.]

I do not know exactly how far the subject which I here propose to consider is akin to that which this great man had in mind; but so far as we can judge from the words [*Analysis Situs*], I am here seeking the first philosophical ground of the possibility of that whose magnitudes Leibniz proposed to determine in a mathematical manner. For the positions of the parts of space in relation to one another presuppose the region towards which they are ordered in such relation; and this region, in ultimate analysis, consists not in the relation of one thing in space to another (which is properly the concept of position) but in the relation of the system of these positions to the absolute world-space. In anything extended the position of parts relatively to one another can be adequately determined from consideration of the thing itself; but the region towards which this ordering of the parts is directed involves reference to the space outside the thing; not, indeed, to points in this wider space—for this would be nothing else but the position of the parts of the thing in an outer relation—but to universal space as a unity of which every extension must be regarded as a part.

As the explanation of these concepts is first to be found in what follows, it is not surprising if the reader as yet finds them very unintelligible; and I therefore limit myself to this one further remark, that my aim in this treatise is to investigate whether there is not to be found in the intuitive judgments of extension, such as are contained in geometry, an evident proof *that absolute space has a reality of its own, independent of the existence of all matter, and indeed as the first ground of the possibility of the compositeness of matter.*

Everybody knows how futile have been the endeavours of the philosophers, by means of the most abstract propositions of metaphysics, to settle this point once for all; and I know of no attempt, save one, to carry this out *a posteriori* as it were, that is, by means of other undeniable propositions which, though themselves lying outside the realm of metaphysics, can afford through their application in the concrete a criterion of their correctness. The one attempt, to which I have referred, was made by the celebrated Euler, the elder, in 1748, as recorded in the *History of the Royal Berlin Academy of Sciences*[1] for that year. But so far from fully achieving his purpose, he only brings to view the difficulty of assigning to the most general laws of motion a determinate meaning, should we assume no other concept of space than that obtained by abstraction from the relation of actual things. The no less notable difficulties which remain in the application of the aforesaid laws, when we endeavour to represent them in the concrete according to the concept of absolute space, are left unconsidered. The proof which I here seek should supply, not to the mechanists (as Herr Euler intended), but to the geometers themselves, a convincing ground for asserting the actuality of their absolute space, and should do so with the evidence to which they are accustomed. With this purpose in view, I make the following preparatory observations.

In physical space, on account of its three dimensions, we can conceive three planes which intersect one another at right angles. Since through the senses we know what is outside us only in so far as it stands in relation to ourselves, it is not surprising that we find in the relation of these intersecting planes to our body the first ground

[1] [Euler there published his *Reflexions sur l'espace et le temps.*]

from which to derive the concept of regions in space.[1] The plane to which the length of our body stands perpendicular is called, in reference to us, horizontal; it gives rise to the distinction of the regions we indicate by *above* and *below*. Two other planes, also intersecting at right angles, can stand perpendicular to this horizontal plane, in such manner that the length of the human body is conceived as lying in the line of their intersection. One of these vertical planes divides the body into two outwardly similar parts and supplies the ground for the distinction between *right* and *left;* the other, which is perpendicular to it, makes it possible for us to have the concept of *before* and *behind*. In a written page, for instance, we have first to note the difference between front and back and to distinguish the top from the bottom of the writing; only then can we proceed to determine the position of the characters from right to left or conversely. Here the parts arranged upon the surface have always the same position relatively to one another, and the parts taken as a whole present always the same outlines howsoever we may turn the sheet. But in our representation of the sheet the distinction of regions is so important, and is so closely bound up with the impression which the visible object makes, that the very same writing becomes unrecognisable when seen in such a way that everything which formerly was from left to right is reversed and is viewed from right to left.

Even our judgments about the cosmic regions are subordinated to the concept we have of regions in general, in so far as they are determined in relation to the sides of

[1] [Kant returned to this subject in 1786 in his treatise, *Was heisst: sich im Denken orientiren?*]

the body. All other relations that we may recognise, in heaven and on earth, independently of this fundamental conception, are only positions of objects relatively to one another. However well I know the order of the cardinal points, I can determine regions according to that order only in so far as I know towards which hand this order proceeds; and the most complete chart of the heavens, however perfectly I might carry the plan in my mind, would not teach me, from a known region, North say, on which side to look for sunrise, unless, in addition to the positions of the stars in relation to one another, this region were also determined through the position of the plan relatively to my hands. Similarly, our geographical knowledge, and even our commonest knowledge of the position of places, would be of no aid to us if we could not, by reference to the sides of our bodies, assign to regions the things so ordered and the whole system of mutually relative positions.

It is even the case that a very notable characteristic of natural organisms, which at times may even give occasion for the distinction of species, consists in the definite direction in which the arrangement of their parts is turned, a feature through which two creatures can be distinguished although they entirely agree both in size and proportion, and even in the position of their parts relatively to one another. The hairs on the crown of every man's head are turned from left to right.[1] The hop-plant always twines round its pole from left to right; beans,

[1] [Cf. Walter Kidd, *The Direction of Hair in Animals and Man*, 1903, pp. 76–77: "Over the posterior fontenelle the familiar whorl or 'crown' is always present and may be to the right or left of the middle line, seldom quite in the middle line, and it is double in a certain number of persons, one whorl lying on each side of the middle line."]

however, take the opposite course. Almost all snails, only some three species excepted, have their spiral turning from left to right, that is, if we proceed from above downwards, from the apex to the mouth.[1] Since in the case of the natural existences just cited the cause of the twist or spiral lies in their very germs, this definite character remains constant in creatures of the same species without any relation to the hemisphere in which they may be found, or to the direction of the daily motion of the sun and the moon, which for us runs from left to right but for our antipodes in the opposite direction. On the other hand, when a certain revolution can be ascribed to the path of these heavenly bodies—as Mariotte[2] professes to have observed in the case of the winds, which from new to full moon tend to work round the whole compass from left to right—this motion must in the other hemisphere go from right to left, as indeed Don Ulloa considers to have been established by his observations in the South Seas.

[1] [The facts, as now known, are more complicated than Kant here suggests. Not only are there many examples among climbing plants both of right-handed (i.e., following the sun, proceeding from left to right) and of left-handed twining stems, but the direction of turning is not always constant throughout a natural order of plants, and sometimes is reversed even in successive internodes of the same stem. In Bittersweet, a plant in our hedgerows, individuals are occasionally found to twine in opposite directions. As regards molluscs, about seventeen genera have spiral shells which are normally sinistral; in about fourteen genera the majority of the species are dextral, but in each genus there are several species with sinistral shells. Some four genera contain species of which the individuals seem to be indifferently either dextral or sinistral. Also there are about two hundred species in which sinistral individuals are exceptionally found among the normal dextral forms. For this information I am indebted to my colleague, Professor J. D. Ashworth.]

[2] [See Kant's treatise on the theory of the winds, *Werke*, Berlin ed., Bd. I, p. 502.]

Since the different feeling of right and left side is of such necessity to the judgment of regions, Nature has directly connected it with the mechanical arrangement of the human body, whereby one side, the right, has an indubitable advantage in dexterity and perhaps also in strength. If, therefore, we set aside individual exceptions which, like cases of squinting, cannot disturb the generality of the rule according to the natural order, all the peoples of the earth are right-handed. In mounting on horseback or striding over a ditch, the body is more easily moved from right to left than vice versa. Everywhere men write with the right hand; with it they do everything for which skill and strength are demanded. But if some investigators, e.g. Borelli and Bonnet, are to be believed, while the right hand seems to have the advantage over the left in mobility, the left has the advantage over the right in sensibility. Borelli likewise assigns to the left eye, and Bonnet to the left ear, the possession of a greater sensibility than the corresponding organ on the right side. And thus the two sides of the human body, in spite of their great outer similarity, are sufficiently distinguished by a well-marked feeling, even if we leave out of account the differing positions of the inner parts and the noticeable beat of the heart, which at every contraction strikes with its apex in oblique motion against the left side of the breast.

What, therefore, we desire to show is that the complete ground of determination of the shape of a body rests not merely upon the position of its parts relatively to one another, but further on a relation to the universal space which geometers postulate—a relation, however, which is such that it cannot itself be immediately perceived. What we do perceive are those differences between bodies which depend exclusively upon the ground which this relation affords. If two figures drawn upon a plane are equal and similar, they can be superimposed. But with physical extension and also with lines and surfaces that do not lie in one plane, the case is often quite different. They can be perfectly equal and similar, yet so different in themselves that the boundaries of the one cannot be at the same time the boundaries of the other. A screw which winds round its axis from left to right will not go into a threaded cylinder whose worm goes from right to left, although the thickness of the stem and the number of turns in an equal length correspond. Two spherical triangles can be perfectly equal and similar, and yet not allow of superposition. But the commonest and clearest example is to be found in the limbs of the human body, which are symmetrically disposed about its vertical plane. The right hand is similar and equal to the left, and if we look at one of them alone by itself, at the proportions and positions of its parts relatively to one another and at the magnitude of the whole, a complete description of it must also hold for the other in every respect.

When a body is perfectly equal and similar to another, and yet cannot be included within the same boundaries, I entitle it the incongruent counterpart of that other. To show its possibility, take a body which is not composed of two halves symmetrically disposed to a single intersecting surface, say a human hand. From all points of its surface draw perpendiculars to a plane set over against it, and produce them just as far behind the plane as these points lie in front of it; the extremities of the lines so produced, if connected, then compose the surface and shape of a physical body which is the incongruent counterpart of the first; i.e., if the given hand is the right, its counterpart is the left. The image of an object in a mirror rests upon the same principle; for it always appears just as far behind the mirror as the object lies in front of its surface, and so the mirrored image of a right hand is always a left. If the object itself consists of two incongruent counterparts, as does the human body when divided by a vertical section from front to back, its image is congruent with it, as can easily be seen by allowing it in thought to make a half turn; for the counterpart of the counterpart of an object is necessarily congruent with the object.

The above considerations may suffice for understanding the possibility of spaces which are completely equal and similar and yet incongruent. We now proceed to the philosophical application of these concepts. From the common example of the two hands, it is already clear that the shape of one body can be completely similar to that of another, and the magnitude of their extension exactly the same, while yet there remains an inner difference, namely that the surface which bounds the one cannot possibly bound the other. Since this surface bounds the physical space of the one but cannot serve as boundary to the other, however one may turn and twist it, this difference must be such as rests upon an inner ground. This inner ground cannot, however, depend on any difference in the mode of connection of the parts of the body relatively to one another; for, as can be seen from the examples adduced, in this respect everything may be completely identical in the two cases. Nevertheless, if we conceive the first created thing to be a human hand, it is necessarily either a right or a left, and to produce the one a different act of the creating cause is required from that whereby its counterpart can come into being.

Should we, then, adopt the conception held by many modern philosophers, especially in Germany, that space consists only in the outer relations of the parts of matter existing alongside one another, in the case before us all actual space would be that which this hand occupies. But since, whether it be right or left, there is no difference in the relations of its parts to one another, the hand would in respect of this characteristic be absolutely indeterminate, i.e., it would fit either side of the human body, which is impossible.

Thus it is evident that instead of the determinations of space following from the positions of the parts of matter relatively to one another, these latter follow from the former. It is also clear that in the constitution of bodies differences are to be found which are real differences, and which are grounded solely in their relation to absolute, primary space. For, only through this relation is the relation of bodily things possible. Since absolute space is not an object of an outer sensation, but a fundamental concept which first makes all such sensations possible, it further follows that whatsoever in the outline of a body exclusively concerns its reference to pure space, can be apprehended only through comparison with other bodies.

A reflective reader will accordingly regard as no mere fiction that concept of space which the geometer has thought out and which clear-thinking philosophers have incorporated into the system of natural philosophy. There is, indeed, no lack of difficulties surrounding this concept, if we attempt to comprehend its reality—a reality which is sufficiently intuitable to inner sense—through ideas of reason. This difficulty always arises when we attempt to philosophise on the first data of our knowledge. But it reaches its maximum when, as in this case, the consequences of an assumed concept [that of spatial relations as subsequent to and dependent on the relations of bodies to one another] contradict the most obvious experience.

DISSERTATION ON THE FORM AND PRINCIPLES OF THE SENSIBLE AND INTELLIGIBLE WORLD

[1770]

DE

MUNDI SENSIBILIS
ATQUE
INTELLIGIBILIS
FORMA ET PRINCIPIIS.

DISSERTATIO PRO LOCO

PROFESSIONIS LOG. ET METAPH. ORDINARIAE RITE SIBI VINDICANDO,

QUAM

EXIGENTIBUS STATUTIS ACADEMICIS

PUBLICE TUEBITUR

IMMANUEL KANT.

RESPONDENTIS MUNERE FUNGETUR

MARCUS HERTZ,

BEROLINENSIS, GENTE IUDAEUS, MEDICINAE ET PHILOSOPHIAE CULTOR,

CONTRA OPPONENTES

GEORGIUM WILHELMUM SCHREIBER,

REG. BOR. ART. STUD.

IOHANNEM AUGUSTUM STEIN,

REG. BOR. I. U. C.

ET

GEORGIUM DANIELEM SCHROETER,

ELBING. S. S. THEOL. C.

IN AUDITORIO MAXIMO

HORIS MATUTINIS ET POMERIDIANIS CONSUETIS

D. XXI. AUG. A. MDCCLXX.

AUGUSTISSIMO

SERENISSIMO ATQUE POTENTISSIMO

PRINCIPI AC DOMINO,

DOMINO

FRIDERICO,

REGI PRUSSORUM,

MARCHIONI BRANDENBURGICO,

S. R. I. ARCHICAMERARIO ET ELECTORI,

SUPREMO SILESIAE DUCI,

ETC. ETC. ETC.

PATRI PATRIAE CLEMENTISSIMO,

REGI AC DOMINO SUO INDULGENTISSIMO,

has demandati sibi muneris primitias devota mente offert

subiectissimus

IMMANUEL KANT.

DISSERTATION ON THE FORM AND PRINCIPLES OF THE SENSIBLE AND INTELLIGIBLE WORLD

SECTION I

ON THE NOTION OF A WORLD IN GENERAL

§ 1. Just as, in dealing with a complex of substances,[1] analysis ends only with a part which is not a whole, i.e., with the *simple;* so synthesis ends only with a whole which is not a part, i.e., with a *world.*[2]

In the following exposition of the concept before us [that of a world], I have had regard not only to the marks which pertain to the distinct knowledge of the object, but also in some measure to its double derivation from the nature of the mind. In this latter regard the exposition may, as an example, serve the end of a deeper insight into metaphysical method, and so seems to me not unworthy of commendation. For it is one thing, given the parts, to conceive the *composition* of the whole by an abstract notion of the intellect, but another thing to take this general notion as a problem set by the reason, and to *follow it out* by the sensitive faculty of apprehension, i.e., to represent it in the concrete by distinct intuition.[3] In the former case composition is secured through the class-concept —in so far as a plurality of things, mutually related, is contained under the concept—and so by ideas which are intellectual and universal. In the latter case the endeavour, which rests on conditions of time, is to attain the con-

[1] [*In composito substantiali.*] [2] [*Mundus.*] [3] [*intuitu.*]

cept of the composite genetically, by the successive addition of part to part, i.e., by *synthesis*, a process which is subject to the laws of *intuition*. Similarly, given a complex of substances, we easily reach the idea of simples by completely removing the intellectual notion of composition; for it is simples that remain when all conjunction is abolished. But according to the laws of intuitive knowledge, the case is different; composition is not removed completely save by a regress from the given whole to all possible parts whatsoever, in other words, by an analysis which again rests on conditions of time.[1] Whereas, then, for the composite [conceptually apprehended] an *aggregate*[2] of parts is required, for a whole [intuitively apprehended] a *totality*[3] is demanded. In consequence neither analysis nor synthesis [the two intuitive processes] will be completed in such a manner that the former will yield the concept of the simple and the latter the concept of the whole, unless each process can be brought to a conclusion in a finite and assignable time.

But since in a *continuous* quantum the regress from the whole to its possible parts, and in an *infinite* quantum the progress from the parts to the given whole, find no end, in the one case the analysis and in the other case the

[1] Each of the words "analysis" and "synthesis" is used in two senses. Synthesis is either qualitative, i.e., progress in series whose members are subordinate to each other, in each case as consequent to ground, or quantitative, i.e., progress in series of co-ordinated members from a given part through its complements to the whole. Similarly, analysis, taken in the first sense, is a regress from consequence to ground, but, in the second sense, it is a regress from a whole to its possible or mediate parts, i.e., the parts of its parts, and so is not the division of but the subdivision of a given compound. Both terms, synthesis and analysis, we here use in the second sense.

[2] [*multitudo.*] [3] [*omnitudo.*]

synthesis will be impossible of completion; the whole cannot, in conformity with the laws of intuition, be apprehended by an exhaustive division of parts, nor the complex by an exhaustive summation. Accordingly, since the unrepresentable and the impossible are commonly regarded as meaning the same thing, and since representation of the concepts of continuity and infinitude in accordance with the laws of intuitive knowledge is clearly impossible, we see how it comes about that these concepts are usually rejected. I am not here pleading the cause of these notions, rejected as they are (especially the concept of continuity) by many schools.[1] It is, however, of the greatest importance that those who follow this highly perverse line of argument should be warned that in so doing

[1] Those who reject the actual mathematical infinite spend no great pains over it. The definition of the infinite which they put together is such that out of it they are able to carve some contradiction. *The infinite* is for them *a quantum than which a greater is impossible;* and the mathematical infinite is an aggregate (of some assignable unit) than which a greater is impossible. But, as they here define not "infinite" but "maximum," and a greatest aggregate is impossible, they easily infer the impossibility of this infinite which they themselves have invented. Or alternatively, they call an infinite aggregate an infinite *number*, and show that this is absurd; and so it clearly is. But again they are fighting only with figments of the mind. If they had conceived the mathematical infinite as a quantum which, in relation to any unit of measurement, is *an aggregate greater than any number;* if, further, they had remarked that measurability here implies reference to the sort of comparison used by the human intellect, one in which any definite concept of an aggregate can be attained only by successive additions of unit to unit, and the complete conception of it which is called number only by completing the summation in a finite time; they would then have clearly seen that things which are at variance with a certain law of a particular kind of [knowing] subject do not thereby pass beyond the sphere of all intellection, since an intellect is possible which might apprehend an aggregate distinctly at a glance, without the successive application of a measure, though such an intellect would not indeed be human.

they fall into a most serious error. Whatever is opposed to the laws of understanding and reason is wholly impossible; but that which (as being an object of pure reason) is at variance only with the laws of intuitive apprehension is not necessarily impossible. For this disagreement between sensibility and intellect (faculties whose character I shall presently expound) shows no more than this, that the mind is frequently unable to follow out in the concrete, and translate into intuitions, abstract ideas which it has received from the intellect. But this subjective inability gives, as very often happens, a false impression of some objective hindrance, and easily deceives the heedless into taking the limits circumscribing the human mind for those within which the very essence of things is contained.

Further, if a complex of substances be given, whether through the testimony of the senses or in any other way, it is easy to see, by an argument based on intellectual grounds, that simples, and a world, are also given. The reasons for this, contained in the very nature of the [knowing] subject, I have indicated in my definition; and I have done so in order that the notion of a world may not appear to be merely arbitrary and, as happens in mathematics, put together only for the purpose of deducing consequences from it. For when the mind is directed upon the conception of the complex, whether by way of breaking it up or of putting it together, it demands and assumes, alike in abstract conception and in actual working out, limits in which it may find rest.

§ 2. The factors[1] to be considered in the definition of a world are these:

[1] [*momenta.*]

I. MATTER, in a transcendental sense, i.e., the parts [of a world] which here are assumed to be *substances*. We might have been quite careless as to the agreement of our definition with the usual signification of the word, since this sort of investigation is made only in view of a problem that arises according to the laws of reason, namely, how it is possible for a number of substances to coalesce into one, and on what condition it depends that this one should not be part of another. But the meaning of the word "world," as it is commonly used, presents itself unsought. For no one reckons accidents as parts of the world, but as determinations of its state. Hence the so-called egoistic world, which is complete in a single simple substance with its accidents, can hardly be fittingly called a world, unless perhaps an imaginary one. For the same reason we cannot refer to the world-whole the series of the successive (i.e., of its states) as a part; for modifications are not parts of a subject but its consequences.[1] Finally, I have not here considered the nature of the substances which make up the world, whether they are contingent or necessary; nor do I needlessly include such determination in the definition, in order thereafter, in the usual way, to elicit it again by some specious process of argument. Later, I shall show that contingency can be abundantly proved from the conditions here posited.

II. FORM, which consists in the co-ordination, not in the subordination, of substances. For co-ordinates are mutually related as complements forming a whole; subordinates as cause and effect, or in general, as principle and consequence.[2] The former relation is reciprocal and homonymous in the sense that each correlate refers to the

[1] [*sed rationata.*]

[2] [*ut principium et principiatum.*]

other as determining it and also as determined by it; the latter is heteronymous, i.e., on the one side it is a relation entirely of dependence, on the other side one entirely of causality. This co-ordination is conceived as real and objective, not as ideal and dependent simply on the choice of the subject, whereby, summing up at pleasure any aggregate whatever, the subject may fashion [for itself] a whole. For by embracing a plurality you may without difficulty make a *whole of representation*, but not, thereby, the *representation of a whole*. Therefore, if perchance there are certain wholes of substances connected with one another by no bond, the mind's encompassing of these, whereby it forces the aggregate into ideal unity, would argue nothing more than a plurality of worlds comprehended by one act of thought. But the bond constituting the *essential* form of a world is regarded as the principle of *possible interactions*[1] of the substances constituting the world. For actual interactions do not belong to essence but to state. The transeunt forces themselves, the causes of the interactions, presuppose some principle through which it is possible that while a plurality of things are, so far as their *subsistence* is concerned, mutually independent of one another, their *states* may mutually refer to one another as grounds and consequents. Apart from such a principle, the assumption of transeunt force in the world cannot be allowed as possible. This form, as being essential to the world, is immutable, and not liable to any alteration. This follows first on logical grounds. Every change presupposes the identity of the subject whose determinations follow one upon another. The world, therefore, as it remains the same world through all its

[1] [*influxuum possibilium.*]

successive states, retains the same fundamental form; identity of the parts does not suffice for identity of the whole; there is required an identity of the characteristic composition. But secondly, it is as following from a real ground that the above principle is mainly established. For the nature of a world, which is the primary, internal, principle of all variable determinations belonging to its state, cannot be opposed to itself, and so is naturally (i.e., by itself) immutable. Thus in any world there is given a certain form which is to be reckoned as belonging to its nature, constant, invariable, as the eternal principle of every contingent transitory form pertaining to the state of the world. Those who think this investigation superfluous are deceived by the concepts of space and time, which seem to be conditions already given *per se*, and primary, and by the aid of which (as they think), without any other principle, it is not merely possible, but necessary, that a number of existing things should be mutually related as joint parts constituting a whole. But, as I shall show later, these notions [of space and time] are certainly not rational, not objective ideas of any real nexus, but phenomena, indicating indeed some common principle of universal connection but not exhibiting it.

III. WHOLENESS,[1] that is, *absolute* totality of conjoint parts. When we have regard to any *given* composite, though it be still part of another, there yet always holds a certain comparative totality, namely, of the parts belonging to that quantum. But in the case of a world, all things whatsoever—whatever things are related together as parts in *any* whole—are understood as being conjointly posited. This absolute totality, though it pre-

[1] [*Universitas*.]

sents itself as an every-day and easily comprehensible concept, especially when it is negatively expressed, as was the case in our original definition,[1] yet when more thoroughly considered is found to confront the philosopher with a crucial problem. For it can hardly be conceived how the never-to-be-completed series of states of the universe, succeeding one another to eternity, can be brought together into a whole comprehending absolutely all changes. On account of its very infinitude it is necessary that it be without limit, and therefore that no series of successive events be given save as part of a further series. Consequently, for the same reason, an all-round completeness, an absolute totality, seems to be ruled out altogether. Though the notion of a part can be taken universally, and though all the things that are contained under this notion, if regarded as placed in the same series, may constitute a unity, yet the concept of a *whole* seems to demand that all those things should be taken together; and this, in the given case, is impossible. For since to the series as a whole there is nothing subsequent, and in the case of a given series of successives the only member to which there is nothing subsequent is the last, there will be to eternity a last member in [the succession]; and this is absurd.

It might be supposed that the difficulty which renders impossible a totality of the successive infinite does not arise in the case of the simultaneous infinite, because simultaneity seems expressly to present itself as including all things in one and the same time. But if a simultaneous infinite be admitted, the totality of the successive infinite must also be conceded; while if the latter is denied, the

[1] [Above, p. 41.]

former must also be given up. For the simultaneous infinite supplies eternity with inexhaustible material to be successively traversed, through its innumerable parts *in infinitum*, and yet this series, freed from all [the bonds of] number, would be supposed to be given in a simultaneous infinite, and so a series which could never be completed by successive addition would be supposed to be capable of being given as a whole.

Anyone who is seeking to escape from this thorny problem should note that the co-ordination of a plurality, whether it be successive or simultaneous (both resting upon concepts of time), does not concern the *intellectual* concept of a whole, but only the conditions of [its] *sensitive intuition;* and so, even if such totalities are not conceivable in sensitive terms, they do not thereby cease to be intellectual concepts. For these it suffices that co-ordinate things be in some way given, and be thought as all belonging to one [whole].[1]

SECTION II

ON THE DISTINCTION OF SENSIBLES AND INTELLIGIBLES IN GENERAL

§ 3. SENSIBILITY[2] is the receptivity of the subject through which it is possible that its power of representation[3] should be affected in a certain manner by the presence of some object. INTELLIGENCE (rationality)[4] is the

[1] [*tanquam pertinentia ad unum.*]

[2] [*Sensualitas.* Kant uses three different adjectival forms, *sensualis, sensibilis,* and *sensitivus;* and these are retained in each case in the translation, cf. § 5 below. While sensual refers to sensation and sensitive refers to the forms of sense (space and time), sensible appears to be employed by Kant in a more general manner as covering both.]

[3] [*status ipsius repraesentativus.*]

[4] [*Intelligentia* (*rationalitas*).]

faculty of the subject through which it is able to represent things which cannot by their own characters[1] act upon the senses. The object of sensibility is the sensible; that which contains nothing save what must be known through intelligence, is the intelligible. The former was called, in the schools of the ancients, phenomenon; the latter, noumenon. Knowledge, so far as it is subject to the laws of sensibility, is sensitive; so far as it is subject to the laws of intelligence, it is intellectual or rational.

§ 4. Thus, on the one hand, all sensitive apprehension depends upon the special nature of the subject, in so far as it is capable of being modified in diverse ways by the presence of objects; and these modifications may differ in different subjects in accordance with variations in the nature of these subjects. But, on the other hand, whatever is exempt from this subjective condition regards only the objects. It is clear, therefore, that things sensitively apprehended are representations of things *as they appear*, while things intellectually known are representations of things *as they are*.

In sense-representation there is in the first place something that we may call *matter*, i.e., sensation; and in addition something that we may call *form*, i.e., that general characteristic of sensible things which makes its appearance in so far as the various things which affect the senses are co-ordinated by a certain natural law of the mind. Further, just as sensation, which constitutes the *matter* of sensual representation, involves indeed the presence of some sensible thing, but for its quality depends upon the nature of the subject, so far as it is modifiable by that object, in the same way, the *form* of the representation indi-

[1] [*per qualitatem suam.*]

cates a certain aspect or relation of the sensa,[1] and yet is not properly an adumbration[2] or schema of the object, but only a certain law inborn in the mind for co-ordinating with one another the sensa arising from the presence of the object. For objects do not strike the senses through their form.[3] In order, therefore, that the various impressions from the object affecting the senses may coalesce into some whole of representation, there is required an internal principle of the mind, through which those various impressions may take on a certain form[4] according to stable and innate laws.

§ 5. To sensual apprehension there thus belong, on the one hand, a matter, which is sensation, and on account of which cognitions are called *sensual;* and, on the other hand, a form,[5] on account of which, even if it be met with empty of all sensation, representations are called *sensitive.* As for the intellectual, on the other hand, it is, above all, to be clearly remarked that the use of the intellect, that is, of the superior faculty of the mind, is double. By the first use, the very concepts of objects or of relations are *given,* and this is the *real use;* by the second use, concepts, whencesoever given, are only *subordinated* to one another, the lower to the higher (the common marks), and compared with one another according to the principle of contradiction; and this is called the *logical use.* The logical use of the intellect is common to all sciences, but not the real use. For a cognition, however given, is regarded as contained under a mark common to many, or as different therefrom; and this either immedi-

[1] [*sensa* is Kant's own term.]

[2] [adumbratio.]

[3] [per formam seu speciem.] [4] [*speciem.*] [5] [*forma.*]

ately and directly (as in judgments, which aim at distinct knowledge) or mediately (as in reasonings, which aim at adequate knowledge).[1] Thus, sensitive cognitions being given, the cognitions are subordinated by the logical use of the intellect to other sensitive cognitions as to common concepts, and as phenomena to more general laws of phenomena.

But here it is of the greatest moment to note that these cognitions, no matter to what extent the logical use of intellect has been exercised upon them, are still to be considered sensitive. For they are called sensitive on account of their origin, not on account of any comparison as to identity or difference. Hence the most *general* empirical laws are none the less sensual. Similarly, those principles of sensitive form (i.e., of determinate relation in space), which are found in geometry, however much the intellect concerns itself with them, arguing according to logical rules from what is sensitively given through pure intuition, do not pass out of the class of sensitives. In things sensual and in phenomena, that which precedes the logical use of intellect is called appearance,[2] and the reflective cognition which arises from the intellectual comparison of a number of appearances is called experience.[3] Thus the only path from appearance to experience is by reflection according to the logical use of intellect. The common concepts of experience are called empirical, and its objects phenomena; the laws of experience, and of all sensitive knowledge in general, are called laws of phenomena. Thus empirical concepts do not become

[1] [Cf. Kant's *Mistaken Subtlety of the Four Syllogistic Figures* (1762), *Werke* (Berlin ed.), II, 58–59; or Abbott's translation, pp. 92–93.]

[2] [*apparentia*.]

[3] [*experientia*.]

intellectual *in the real sense*[1] by being brought to a greater universality, and so pass out of the class of sensitive knowledge. However high they ascend by way of abstraction, they always remain sensitive.

§ 6. When we come to the objects of intelligence which are strictly such,[2] in which the use of the intellect is *real*, the concepts involved, whether of objects or relations, are given through the very nature of the intellect, not abstracted from any use of the senses, and do not contain any form of sensitive cognition as such. But here it is necessary to take notice of a very great ambiguity in the word "abstract," which had best be disposed of at once, that it may not vitiate our treatment of the intellectual. Properly, we should say that we *abstract from* something, not that we *abstract* something. The first signifies that in a certain concept we do not attend to other things connected in some way with it. The second however means that it is not given save in the concrete, and so has to be separated from what is conjoined with it. Hence an intellectual concept is not abstracted from the sensitive, but abstracts from all that is sensitive, and perhaps would be more correctly called *abstracting* than *abstracted*. It is therefore more appropriate to call intellectual ideas *pure*, and concepts given only empirically *abstract*.

§ 7. From this it can be seen that it is wrong to regard the sensitive as that which is more confusedly known, and the intellectual as being that of which our knowledge is distinct. For these are merely logical distinctions, and plainly do not affect those data which un-

[1] [*in sensu reali*, i.e., as distinguished from the formal or *logical* sense.]

[2] [*intellectualia stricte talia*.]

derlie all logical comparison. For that matter, the sensitive may be very distinct, and the intellectual extremely confused. This is shown, on the one hand by geometry, the prototype of all sensitive knowledge, and on the other hand by metaphysics, the instrument in all things [strictly] intellectual. Everyone knows how much labour metaphysics devotes to dispersing the clouds of confusion which darken the common intellect, though its work has not always as happy an issue as that of geometry. None the less, each kind of knowledge preserves the mark of its descent, so that the former kind, however distinct, is, on account of its origin, called sensitive; the latter kind, however confused, remains intellectual. Moral concepts, for instance, though of this confused character, are known not by experience but by the pure intellect itself. But I fear that the illustrious Wolff by means of this distinction, which for him is merely logical, between the sensitive and the intellectual may (to the great detriment of philosophy) have quite destroyed that noblest enterprise of antiquity, the determining of the nature of phenomena and of noumena, turning men's minds from these investigations to what are frequently but logical minutiae.

§ 8. That part of philosophy which contains the first principles of the use of pure intellect is metaphysics. Introductory to it is the science which brings out the distinction between sensitive and intellectual knowledge; and of this science the present dissertation is a specimen. Since, then, no empirical principles are to be found in metaphysics, the concepts there met with are not to be looked for in the senses, but in the very nature of pure intellect, not as concepts *connate* to it, but as abstracted (by attention to its actions on the occasion of experience) from laws inborn in the mind, and so to this extent as *acquired.* Concepts of this sort are: possibility, existence, necessity, substance, cause, etc., with their opposites or correlates. These never enter into any sensual representation as parts of it, and could not, therefore, in any wise be abstracted therefrom.

§ 9. The concepts of understanding have, in especial, two functions: (*a*) In their *critical*[1] use, they perform the negative service of keeping sensitive concepts from being applied to noumena. Though they advance knowledge not at all, they yet keep it free from the contagion of errors. (*b*) In their *dogmatic*[2] use, the general principles of pure understanding, such as are dealt with in ontology or rational psychology, issue in some exemplar, which is conceivable only by pure intellect, and is the common measure of all other things as far as real. This exemplar—*Perfectio Noumenon*—is perfection either in a theoretical[3] or in a practical sense. In the former, it is the Supreme Being, God; in the latter, moral perfection. Thus moral philosophy, so far as it supplies first principles of moral judgment, is known only through the pure understanding, and itself belongs to pure philosophy. Epicurus, who reduced the criteria of morals to the feeling of pleasure or unpleasantness, is therefore quite rightly condemned, along with certain moderns who, like Shaftesbury and his school, follow him, in a much less thorough manner, part

[1] [*elencticus.*]

[2] [*dogmaticus.*]

[3] We regard a thing theoretically so far as we attend only to what pertains to existence; but practically if we consider what, through freedom, ought to be in it.

of the way. In every kind of existence in which quantity is variable the maximum is the common measure and principle whereby we have knowledge. The maximum of perfection, which is called by Plato an idea (as in his "idea" of the state), we now entitle an ideal. It is the principle of all that is contained under the general notion of any perfection, in so far as smaller degrees are supposed not to be determinable save by limiting the maximum. God, however, while as ideal of perfection he is the principle of knowledge, is at the same time, as really existing, the principle of the coming into existence of all perfection whatsoever.

§ 10. No intuition of things intellectual, but only a symbolic knowledge of them, is given to man. Intellection is possible to us only through universal concepts in the abstract, not through a singular concept in the concrete. For all our intuition is bound to a certain formal principle under which alone anything can be apprehended by the mind immediately, that is, as *singular*, and not as merely conceived discursively through general concepts. But this formal principle of our intuition (space and time) is the condition under which anything can be an object of our senses; and being thus the condition of sensitive knowledge, it is not a means of *intellectual* intuition. Further, all the matter of our knowledge is given by the senses alone, whereas a noumenon, as such, is not to be conceived through representations derived from sensations. Consequently, an intelligible concept is, as such, destitute of all that is given by human intuition. Thus, in our minds intuition is always passive, and so is possible only so far as something is able to affect our senses. But the divine intuition, which is the ground of its objects, not consequent on them,[1] is, owing to its independence, archetypal, and so is completely intellectual.

§ 11. But although phenomena are, properly, appearances,[2] not ideas,[3] of things, and express no internal and absolute quality of the objects, the knowledge of them is none the less quite genuine knowledge. For, in the first place, so far as they are sensual concepts or apprehensions, they bear witness, as being caused, to the presence of an object—which is opposed to idealism.[4] On the other hand, to take judgments about what is known by sense, the truth of a judgment consists in the agreement of its predicate with the given subject. But the concept of the subject, so far as it is a phenomenon, can be given only by its relation to the sensitive faculty of knowledge; and it is by the same faculty that sensitively observable predicates are also given. Hence it is clear that the representations of subject and of predicate arise according to common laws, and so allow of a perfectly true knowledge.

§ 12. All things which, as objects, are referred to our senses, are phenomena; whatever does not affect the senses but contains merely the special form of sensibility, belongs to pure intuition (i.e., to intuition empty of sensations, but not for that reason intellectual). The phenomena of outer sense are reviewed and expounded in physics, those of inner sense in empirical psychology. But pure intuition (in man) is not a universal or logical concept *under* which, but a singular concept *in* which, all sensibles are apprehended. That is to say, it contains the

[1] [*qui objectorum principium, non principiatum.*]

[2] [*species.*] [3] [*ideae.*]

[4] [Kant in 1770, as in certain passages in the *Critique of Pure Reason*, appears to have regarded idealism as virtually teaching what may be called illusionism.]

concepts of space and time. Since these in no way determine the quality of sensibles, they are objects of science only as determining things according to quantity. Pure mathematics considers space in geometry, time in pure mechanics. To these there is added a certain concept which, though itself indeed intellectual, yet demands for its actualization in the concrete the auxiliary notions of time and space (in the successive addition and simultaneous juxtaposition of a plurality), namely, the concept of number, treated of by arithmetic. Thus pure mathematics, in that it deals with the form of all our sensitive knowledge, is the organon of all knowledge which is at once intuitive and distinct; and since its objects are not merely formal principles of all intuition, but themselves original intuitions, it yields us quite genuine knowledge, and at the same time furnishes a model of the highest certainty for knowledge in other fields. There is thus a science of things sensual, although, since they are phenomena, the use of the understanding in reference to them is not real but only logical. From this it is clear in what sense those thinkers who derive their inspiration from the Eleatic School are to be understood as having denied scientific knowledge of phenomena.

SECTION III

ON THE PRINCIPLES OF THE FORM OF THE SENSIBLE WORLD

§ 13. A principle of the form of the universe is one which contains the ground of a universal connection,[1] whereby all substances and their states belong to one and the same whole, that is, to a world. A principle of the

[1] [*nexus.*]

form of the sensible world is one which contains the ground of a universal connection of all things so far as they are phenomena. A form of the intelligible world implies an objective principle, that is, some cause in virtue of which there is a connection between things existing in themselves. But the world, regarded as phenomenon, that is, in relation to the sensibility of the human mind, acknowledges no principle of its form save a subjective one, that is to say, a law of the mind, on account of which all things which can (through their quality) be objects of the senses must necessarily be presented as belonging to the same whole. Thus whatever principle of the form of the sensible world may finally be acknowledged, it yet includes within its range only such existences[1] as are deemed possible objects of sense, and so extends neither to immaterial substances (which, as such, are by definition at once excluded altogether from the outer senses) nor to the cause of the world (which cannot be an object of the senses, since through it the mind itself exists and is endowed with the power of sense). I shall now show that there are two such formal principles of the phenomenal universe which are absolutely primary and universal,[2] and which are, as it were, the schemata and conditions of all human knowledge that is sensitive. I refer to time and space.

§ 14. On time.

1. *The idea of time does not originate in the senses, but is presupposed by them.* For the impressions of sense[3] can be represented either as simultaneous or as successive only

[1] [*actualia.*]

[2] [*catholica.*]

[3] [*Quae enim in sensus incurrunt.*]

through the idea of time; succession does not beget the concept of time, but presupposes it. Thus the notion of time (regarded as acquired through experience) is very badly defined in terms of the series of actual things existing *after* one another. For what the word *after* may signify, I know only by means of an antecedently formed concept of time. Things are one after another when they exist at *different times*, just as things are simultaneous when they exist at the *same time*.

2. *The idea of time is singular, not general*. For no time is apprehended except as part of one and the same boundless[1] time. If we think of two years we cannot represent them save by a determinate dating with regard to one another, and if they do not follow one another immediately, save as joined to one another by some intermediate time. But which of different times is earlier, which later, can by no means be defined by any marks conceivable by the intellect, unless we are to incur a vicious circle. The mind does not distinguish earlier and later except by a *singular* intuition. Further, we conceive all actual things as located *in* time, not as contained *under* its general notion as under a common mark.

3. *The idea of time is therefore an intuition*. And since the idea is conceived prior to all sensation, as a condition of relations exhibited in sensibles, the intuition is not sensual but *pure*.

4. *Time is a continuous quantum*. It is the principle of the laws of continuity in the changes of the universe. For a quantum is continuous which is not composed of simples. But since through time there are apprehended relations alone, without any given things related to one an-

[1] [*immensi*.]

other, it follows that in time as a quantum we have a complex of such a kind that when the complexity is thought away nothing remains. But any complex of which nothing at all remains if its composition is wholly removed, does not consist of simple parts; therefore, etc. Thus any part of time is a time, and the simples which exist in time, namely, moments, are not parts of time, but limits[1] between which there is a stretch of time. For, given two moments, a time is not given except in so far as in these moments an actual succession takes place. Therefore, besides a given moment it is necessary that there be given a time in whose later part there is another moment.

The metaphysical law of continuity is this: all changes are continuous, or flow; i.e., opposite states do not succeed one another except through an intermediate series of different states. For since two opposite states are in different moments of time, and between two moments of time there is always some intervening time, and in the infinite series of its moments the substance is not in either of the given states, nor yet in no state, it will be in different states, and so *in infinitum*.

The celebrated Kaestner, in proceeding to examine this Leibnizian law, challenges its defenders[2] to prove that the continuous motion of a point along all the sides of a triangle is impossible, a proposition which assuredly requires to be proved if the law of continuity is granted. Here, then, is the desired demonstration. Let the letters $a\ b\ c$ denote the three angular points of a rectilinear triangle. If a moving point traverses with a continuous mo-

[1] [*termini.*]

[2] *Höhere Mechanick* [Göttingen, 1766], S. 354.

tion the lines *ab*, *bc*, *ca*, i.e., the whole perimeter of the figure, it is necessary that it move through the point *b* in the direction *ab*, and through the same point *b* in the direction *bc*. But since these motions are diverse, they cannot take place simultaneously. Therefore the moment of the presence of the moving point in the vertex *b*, so far as it moves in the direction *ab*, is different from the moment of the presence of the moving point in the same vertex *b*, so far as it moves in the direction *bc*. But between two moments there is a time; therefore the moving point is present during some time in the same point, i.e., is at rest, and thus does not proceed with a continuous motion, which contradicts the hypothesis. The same demonstration holds of motion along any straight lines including an assignable angle. Therefore a body in continuous motion changes its direction only according to a line no part of which is straight, i.e., in a curve, as Leibniz declared.

5. *Time is not something objective and real.* It is neither substance nor accident nor relation, but is a subjective condition, necessary owing to the nature of the human mind, of the co-ordinating of all sensibles according to a fixed law;[1] and it is a *pure intuition.* For we co-ordinate alike substances and accidents, whether according to simultaneity or according to succession, only through the concept of time; and thus the notion of time, as a formal principle, is prior to the concepts of simultaneity and succession. As for relations[2] of whatever sort, in so far as they come within the scope of the senses (namely, as to their simultaneity or succession), they involve nothing

[1] [sed subjectiva condicio per naturam mentis humanae necessaria, quaelibet sensibilia certa lege sibi coordinandi.]

[2] [relationes s. respectus.]

further [than is involved in the co-ordination of substances and accidents] except this, the determining of positions in time, as either in the same point of it, or in different points.

Those who assert the objective reality of time, conceive it in one or other of two ways. Among English philosophers especially, it is regarded as a continuous real flux, and yet as apart from any existing thing—a most egregious fiction. Leibniz and his School declare it to be a real characteristic, abstracted from the succession of internal states. This latter view at once shows itself erroneous by involving a vicious circle in the definition of time, and also by entirely neglecting simultaneity,[1] a most important consequence of time. It thus upsets the whole use of sane reason, in as much as, instead of requiring the laws of motion to be defined in terms of time, it would have time itself defined in respect of its own nature by reference to the observation of moving things or of some series of internal changes—a procedure by which, clearly, all the certainty of our rules is lost. But as for the fact that we cannot estimate *quantity* of time save in the

[1] What is simultaneous is not made simultaneous simply by not being successive. For when succession is taken away there is indeed removed a certain conjunction of things within the temporal series, but there does not on that account at once arise another real relation, such as is the conjunction of all in the same moment. For simultaneous things are joined in the same moment of time just as successive things in different moments. Thus though time possesses only one dimension, yet the ubiquity of time (to use Newton's manner of speaking), owing to which all things conceivable by sense are *at some time*, adds to the quantum of actuals a second dimension, so far as they hang, as it were, from the same point of time. For if you represent time by a straight line produced to infinity, and simultaneous things at any point of time by lines drawn perpendicular to it, the plane thus generated will represent the phenomenal world, both as to its substance and as to its accidents.

concrete, namely, either by motion or by the series of [our] thoughts, this is because the concept of time rests only on an internal law of the mind. For since the concept is not a connate intuition, the action of the mind, in the co-ordinating of its sensa, is called forth only by the help of the senses. So far is it from being possible that anyone should ever deduce and explain the concept of time by the help of reason, that the very principle of contradiction presupposes it, involving it as a condition. For A and not-A are not incompatible unless they are judged of the same thing *together* (i.e., in the same time); but when they are judged of a thing successively (i.e., at different times), they may both belong to it. Hence the possibility of changes is thinkable only in time; time is not thinkable through changes, but *vice versa*.

6. But though time, posited in itself and absolutely, is an imaginary being, yet so far as it is related to an immutable law of sensibles as such, it is a quite genuine concept, and a condition of intuitive representation, extending *in infinitum* through all possible objects of the senses. For since simultaneous things, as such, cannot be presented to the senses except with the help of time, and changes are thinkable only through time, it is clear that this concept contains the universal form of phenomena, and accordingly that all events observable in the world, all motions and all internal changes, necessarily agree with any axioms, such as we have partly expounded, which can be determined in regard to time, since only under these conditions can objects of the senses be, and be co-ordinated. It is therefore absurd to wish to incite reason against the first postulates of pure time (e.g., continuity, etc.), since these follow from laws than which

nothing more primary or original can be found. Reason itself cannot, in the use of the principle of contradiction, dispense with the support of this concept of time, so primitive and original is it.

7. Thus time is an absolutely primary, formal principle of the sensible world. For all things that are in any way sensibles can be apprehended only as at the same time or in successive times, and so as included and definitely related to each other within the course of the one single time. Thus through this concept, primary in the domain of sense, there necessarily arises a formal whole which is not a part of any other, i.e., the phenomenal world.

§ 15. On space.

A. *The concept of space is not abstracted from outer sensations.* For I cannot conceive anything as located outside me unless I represent it as in a space different from the space in which I myself am; nor can I conceive things as outside one another unless I arrange them in different parts of space. Therefore the possibility of outer perceptions, as such, presupposes, and does not create, the concept of space. Inasmuch as, moreover, things which are in space [and which presuppose space] affect the senses, space itself [on which the apprehension of them depends] cannot be derived from the senses.

B. *The concept of space is a singular representation*, including all spaces *in* itself, not an abstract common notion containing them *under* itself. For what we call "many spaces" are only parts of the same boundless space, with a certain position relatively to one another; nor can we conceive a cubic foot except as co-terminous on all sides with surrounding space.

C. *The concept of space is thus a pure intuition*, since it is a singular concept. It is not put together from sensations, but is the fundamental form of all outer sensation. This pure intuition can be readily observed in the axioms of geometry, and in every mental construction of postulates or of problems. For that space has not more than three dimensions, that there is but one straight line between two points, that from a given point in a plane surface with a given straight line as radius a circle can be described, etc., are not inferred from any universal notion of space, but can only be discerned[1] in space in the concrete. We cannot by any sharpness of intellect describe discursively, that is, by intellectual marks, the distinction in a given space between things which lie towards one quarter, and things which are turned towards the opposite quarter. Thus if we take solids completely equal and similar but incongruent, such as the right and left hands (so far as they are conceived only according to extension), or spherical triangles from two opposite hemispheres, although in every respect which admits of being stated in terms intelligible to the mind through a verbal description they can be substituted for one another, there is yet a diversity which makes it impossible for the boundaries of extension to coincide. It is therefore clear that in these cases the diversity, that is, the incongruence, cannot be apprehended except by pure intuition. Hence geometry employs principles which not only are unquestioned and discursive, but which are such as fall under the mind's direct observation. *Evidence* in demonstrations (meaning thereby the *clearness* of assured knowledge, so far as this clearness can be likened to that of sense) is found in

[1] [*cerni.*]

geometry in not merely in the highest degree, but is found there alone of all the pure sciences. Geometrical evidence is thus the model for, and the means of attaining, all evidence in the other sciences. For since geometry contemplates the relations of space, the concept of which contains in itself the very form of all sensual intuition, there can be nothing clear and perspicuous in things perceived by outer sense except through the mediation of the intuition which that science is occupied in contemplating. Further, geometry does not demonstrate its universal propositions by apprehending the object through a universal concept, as is done in matters of reason, but by submitting it to the eyes in a singular intuition, as is done in matters of sense.[1]

D. *Space is not something objective and real*, neither substance, nor accident, nor relation, *but subjective and ideal;* and, as it were, a schema, issuing by a constant law from the nature of the mind, for the co-ordinating of all outer sensa whatsoever. Those who defend the reality of space, either conceive it as an absolute and boundless receptacle of possible things (the view commends itself to most geometers, following the English), or hold that it is itself a relation of existent things, vanishing therefore if things be annihilated, and not thinkable except in actual

[1] I here pass over the proposition that space must necessarily be conceived as a continuous quantum, since that is easily demonstrated. Owing to this continuity it follows that the simple in space is not a part but a limit. But a limit in general is that in a continuous quantum which contains the ground of its limits [*limes;* elsewhere, throughout, the term used is *terminus*]. A space which is not the limit of a second space is complete, i.e., a solid. The limit of a solid is a surface, of a surface a line, of a line a point. Thus there are three sorts of limits in space, just as there are three dimensions. Of these limits two, surface and line, are themselves spaces. The concept of limit has no application to quanta other than space and time.

things (as, following Leibniz, most of our countrymen maintain). The former empty figment of reason, since it imagines an infinity of real relations without any things which are so related, pertains to the world of fable. But those who adopt the second opinion fall into a much more serious error. For while the former party only place a stumbling block in the way of certain concepts which are rational (i.e., concern noumena), and which are in any case of highly abstruse character (e.g., questions concerning a spiritual world, omnipresence, etc.), the latter set themselves in direct opposition to the actual phenomena, and to geometry, that most faithful interpreter of all phenomena. For—to pass over the obvious circle in which they are necessarily involved in their definition of space—they dash down geometry from the supreme height of certainty, reducing it to the rank of those sciences whose principles are empirical. For if all properties of space are borrowed only from external relations through experience, geometrical axioms do not possess universality, but only that comparative universality which is acquired through induction and holds only so widely as it is observed; nor do they possess necessity, except such as depends on fixed laws of nature; nor have they any precision save such as is matter of arbitrary convention; and we might hope, as in empirical matters, some day to discover a space endowed with other primary affections, and perhaps even a rectilinear figure enclosed by two straight lines.

E. Although the concept of space, viewed as an objective and real being or affection, is imaginary, nevertheless relatively to all sensibles it is not merely altogether true, but the foundation of all truth in outer sensibility.

For things cannot appear to the senses in any manner[1] except by the mediating power of the mind, co-ordinating all sensations according to a constant law inborn in it. Nothing whatsoever, then, can be given to the senses save in conformity with the primary axioms of space and the other consequences of its nature, as expounded by geometry. Though the ground or principle of these axioms be only subjective, it will necessarily be in harmony with them, because only so does it harmonise with itself. The laws of sensibility will be laws of nature, in so far as nature is able to affect the senses. Thus as regards all properties of space which are demonstrated from a hypothesis not invented, but intuitively given as being a subjective condition of all phenomena, nature is meticulously conformed to the rules of geometry, and only in accordance with them can nature be revealed to the senses. Certainly, unless the concept of space had been given originally through the nature of the mind (so that anyone who should labour to imagine any other relations than are anticipated in it would waste his pains, since he would have been forced to use this very concept in support of his fiction), the use of geometry in natural philosophy would be very unsafe; for it would be possible to doubt whether the notion of space obtained from experience will sufficiently harmonise with nature, the determinations from which it has been abstracted being perchance denied—a suspicion which has indeed occurred to some thinkers. Space, therefore, is an absolutely first formal principle of the sensible world, not only for the reason that the objects composing the universe cannot be phenomena save through the concept of space, but especially for this reason, that, by its

[1] [*sub ulla specie.*]

essence, it must necessarily be single, embracing absolutely all outer sensibles, and so constitutes a principle of totality,[1] i.e., of a whole which cannot be part of another whole.

Corollary. These, then, are the two principles of sensitive knowledge, not general concepts (as is the case in matters of intellect), but singular intuitions which yet are pure. In these intuitions it is not true that the parts, and in especial the simple parts, contain, as the laws of reason prescribe, the ground of the possibility of the composite; but instead, after the pattern of sensitive intuition, the infinite contains the ground of every thinkable part, and finally of the simple, or rather of the limit. For only if infinite space and time be given, can any definite space or time be marked out by limitation of it; neither a point nor a moment can be thought by itself; they are conceived only as limits in an already given space or time. Thus all primary properties of these concepts are beyond the jurisdiction of reason, and so cannot in any way be intellectually explained. But none the less they are the pre-suppositions upon which the intellect rests when, with the greatest possible certainty, and in accordance with logical laws, it draws consequences from the primary data of intuition.

Of these concepts, the one [i.e., space] properly concerns the intuition of an *object*, the other [i.e., time] a *state*, namely, that of representation. Thus space is applied as an image[2] to the concept of time itself, representing it by a line, and its limits (moments) by points. But time approaches more nearly to a universal, rational concept, in that it embraces absolutely everything within its

[1] [*universitatis*.] [2] [*typus*.]

survey, namely, space itself, and in addition the accidents which are not comprehended in space-relations, such as the thoughts of the soul. Further, though time does not indeed prescribe laws to reason, it yet establishes the chief conditions by the help of which the mind can order its notions according to the laws of reason. Thus I cannot decide whether a thing is impossible, except by predicating A and not-A of the same subject *at the same time.* Above all, when the intellect is applied to experience, though the relation of cause and effect, even in outer objects, involves *space* relations, none the less in all cases, outer and inner alike, the mind can be informed what is earlier and what later (i.e., which is cause and which effect) only by means of the time relation. Indeed even the quality of space cannot be made intelligible, unless we relate it to some measure as unit and express it by a number, which is itself only an aggregate distinctly apprehended by numeration, i.e., by the process of adding one to one successively in a given time.

Finally, the question naturally arises whether these concepts are connate or acquired. The latter alternative, it is true, seems already refuted by our demonstrations, but the former is not to be rashly admitted, since, in appealing to a first cause, it opens a path for that lazy philosophy which declares all further research to be vain. Both concepts are without doubt acquired, as abstracted, not indeed from the sensing of objects[1] (for sensation gives the matter, not the form, of human apprehension), but from the action of the mind in co-ordinating its sensa according to unchanging[2] laws—each being, as it were, an immutable type, and therefore to be known intuitively.

[2] [*perpetuas.*] [1] [*a sensu objectorum.*]

For, though sensations excite this act of the mind, they do not determine the intuition. Nothing is here connate save the law of the mind, according to which it combines in a fixed manner the sensa produced in it by the presence of the object.

SECTION IV

ON THE PRINCIPLE OF THE FORM OF THE INTELLIGIBLE WORLD

§ 16. Those who regard space and time as a real and absolutely necessary bond, as it were, of all possible substances and states, consider that nothing else is needed for conceiving how a certain original relation can belong to a plurality of existents, as the primary condition of their possible interactions and as the principle of the essential form of the universe. For since, according to their view, all things that exist are necessarily somewhere, it seems to them superfluous to enquire why they are presented in a fixed manner, that being determined directly by the fact that space is a whole which includes all things. Apart, however, from the fact that this concept of space, as already proved, concerns rather the laws of the sensibility of the subject than conditions of the objects themselves, however much reality is assigned to it, it yet denotes only the intuitively given possibility of universal co-ordination. The question, which cannot be solved save by the intellect, thus still remains quite untouched: *on what principle rests this relation of all substances, which, when viewed intuitively, is called space.* Thus the question, as to the principle of the form of the intelligible world, hinges upon its being made clear, how it is possible that a plurality of substances should stand in relations of mutual interaction, and in this way belong to the same whole, called a world. We are here contemplating the world not as to its matter (whether the substances of which it consists are material or immaterial), but as to its form, that is, how in general the connection of many substances, and the totality of them all, is brought about.

§ 17. Given a plurality of substances, a principle of possible interaction between them is not given by their mere existence; something further is required from which their mutual relations may be understood. For through their mere subsistence they do not necessarily refer to anything else, except perhaps to their cause; and the relation of effect to cause is not one of interaction, but of dependence. If, therefore, there be any interaction between some and others, there is required a special ground determining this precisely.

Herein consists the primal error of the *influxus physicus* [doctrine] in its common meaning; it hastily assumes interaction of substances and transeunt forces as sufficiently comprehensible through the mere existence of the substances, and so is not so much a system as the neglect of all philosophic system as unnecessary in dealing with this question. But if we rid the concept of *influxus physicus* of this defect, we have the kind of interaction which alone deserves to be called real, and which also justifies us in speaking of the world as a whole which is real, not ideal and imaginary.

§ 18. A whole of necessary substances is impossible. Since for each its own existence is securely established, apart from all dependence (which clearly is not appropriate to things necessary) on anything else, it is evident not only that the interaction of substances (i.e., the reciprocal dependence of their states) does not follow from their existence, but also that to them as necessary beings it cannot be attributed at all.

§ 19. Thus a whole of substances is a whole of contingent things, and the world, by its essence, consists of merely contingent things. Further, no necessary substance is connected with the world, save as cause is connected with effect, and not, therefore, as a part is connected with its complementary parts to form a whole. For the connection of joint-parts is that of mutual dependence, which does not apply to a necessary being. Thus the cause of the world is an extramundane being; consequently it is not the soul of the world;[1] its presence in the world is not local but virtual.[2]

§ 20. The substances of the world are beings deriving from another, not, however, from different beings, but all from one. For suppose them to be the effects of a plurality of necessary beings; then, since the causes are without any mutual relation, the effects would not be in interaction. Accordingly unity in the conjunction of the substances of the universe is a consequence of the dependence of all upon one. The form of the universe bears witness to the cause of its matter, and only a single cause of all things can be the cause of its totality; there cannot be an architect of the world who is not also its creator.

§ 21. If there were a plurality of first necessary causes with their effects, these would be worlds, not a world, because they would not be in any way connected to form the same whole. And *vice versa*, if there be a plurality of actual worlds outside one another, there is a plurality of first and necessary causes, with no interaction between

[1] [*anima mundi.*] [2] [I.e., dynamical: a thing is where it acts.]

one world and another, nor between the cause of one world and the world depending upon another cause.

It follows that a plurality of actual worlds outside one another is impossible not by its very concept (as Wolff[1] wrongly concluded from the notion of a complex, or aggregate, a notion which he thought equivalent to a whole as such), but only under this condition, that there exists one single necessary cause of all things. If more than one be admitted, a plurality of worlds (in the strictest metaphysical sense), external to one another, will be possible.

§ 22. We can conclude from a given world to a sole cause of all its parts. If we could similarly argue, *vice versa*, from a given cause common to all things to their interconnection, and so to the form of a world (though I confess this conclusion does not seem to me equally clear), then owing to the substances all being maintained by a common principle, the fundamental connection of the substances would not be contingent but necessary. The harmony arising from their very subsistence, resting upon this common cause, would proceed according to common rules. This kind of harmony I name generally established harmony. That which occurs only so far as each special state of a substance is adapted to the state of another substance, I entitle specially[2] established harmony. According to the former, interaction is real and physical; according to the latter ideal and sympathetic. Thus all interaction of the substances of the universe is *externally* established, through the cause common to all; and this

[1] [The following references to Wolff's works are given in the Berlin edition of Kant's *Werke*, II, p. 512. *Cosmologia generalis* (edition of 1737), §§ 48, 60, 61; *Vernünftige Gedancken* (8th ed., 1741), §§ 548–50, 948–50.]

[2] [*singulariter*.]

is either generally established through *influxus physicus* (in an emended sense[1]), or is brought about for the states of each substance individually. In the latter case, it is either originally founded in the primary constitution of every substance (pre-established harmony), or impressed on the occasion of every change (occasionalism). If, then, owing to the maintenance of all substances by a unitary ground, there be a necessary conjunction of all things whereby they constitute a unity, the universal interaction of substances will be by *influxus physicus*, and the world will be a real whole; if otherwise, the interaction will be sympathetic (i.e., harmony without real interaction), and the world only an ideal whole. In my view, though the former alternative has not been demonstrated, there are otherwise ample reasons for its acceptance.

Scholium.

If it were legitimate to overstep a little the limits of apodeictic certainty befitting metaphysics, it might be worth while to investigate certain questions concerning not merely those laws but also those causes of sensitive intuition which can only be known through the intellect. The human mind, we might then say, is affected by outer things, and the world lies open to its view *in infinitum*, only in so far as the mind, along with all other things, is upheld by the infinite power of a single cause. For this reason it senses external things only through the presence of the one upholding common cause; and space, which is the universal and necessary condition, sensitively apprehended, of the co-presence of all things, can therefore be entitled *omnipraesentia phaenomenon*. (For it is not be-

[1] [cf. above, § 17]

cause the cause of the universe is in the same place with each and everything that it is present to them all; on the contrary, places, i.e., possible relations of substances, exist, because it, the cause, is inwardly present to all things). Again, the possibility of all changes and successions—the principle of which, so far as it is sensitively known, resides in the concept of time—presupposes the persistence of a subject whose opposite states succeed one another. But that whose states change does not persist unless maintained by another; and thus the concept of time as single, infinite, and immutable,[1] in which are and persist all things, is the *aeternitas phaenomenon* of the general cause. But it seems wiser to hug the shore of the knowledge granted us by the mediocrity of our intellect, than thus, after the manner of Malebranche, whose opinion is little different from that just expounded, namely, that we see all things in God, to push out into the open sea of mystical enquiries.

SECTION V

ON THE METHOD OF DEALING WITH THE SENSITIVE AND THE INTELLECTUAL IN METAPHYSICS

§ 23. In all sciences whose principles are given intuitively, whether by sensual intuition (experience) or by intuition which while indeed sensitive is pure (the concepts of space, of time and of number), that is to say, in mathematics and natural sciences, *practice gives rise to method.*

[1] The moments of time do not present themselves as successive to one another; for, in that case another time would have to be presupposed for the succession of the moments; but in sensitive intuition actual things seem to descend, as it were, through a continuous series of moments.

After a science, by trial and discovery, has reached some degree of comprehensiveness and order, we are able to see by what method we must proceed to carry it to completion, clearing it of the stains of error and confused thought, and so bringing it to its highest purity. It was in this way that Grammar, after abundant use of speech, provided an opportunity for rules and training; and similarly with style, after high achievement in poetry and oratory. In the case of all sciences whose primary concepts and axioms are thus given by sensitive intuition, the use of the intellect is only logical, i.e., only applied to the mutual subordination of cognitions in respect of their universality, conformably to the principle of contradiction, for instance, phenomena to more general phenomena, the corollaries of pure intuition to the intuitive axioms.

In pure philosophy, however, such as metaphysics, in which the use of the intellect in dealing with principles is real, that is to say, in which primary concepts of things and relations, and axioms themselves, are primarily given by the pure intellect itself, and which, since they are not intuitions, are not immune from error, *method precedes all science.* Consequently, all attempts at philosophizing made antecedently to the thorough investigation and firm establishment of its rules seem to be rashly conceived, and ought to be set aside as among the idle triflings of the mind. For since the right use of reason here establishes the principles themselves, and since only by virtue of its own powers do the objects and the axioms which are to be thought about them first become known, the exposition of the laws of pure reason is the very beginning of the science, and the distinction of these laws from spurious ones is the criterion of truth. At present the method of this science is not well known; except so far as logic prescribes a method to all sciences in general, that which is adapted to the special nature of metaphysics is completely ignored. That those devoted to this enquiry seem hitherto, with all the endless rolling of their Sisyphean stone, to have accomplished scarcely anything, is not, therefore, surprising. I have neither the mind nor the space to speak here at length on so notable and far-reaching a question, but I shall briefly expound one point which constitutes an important part of this method, namely, *the contamination of intellectual knowledge by the sensitive*. This contamination not only misleads the unwary in the application of principles, but even introduces spurious principles in the guise of axioms.

§ 24. The method of all metaphysics in dealing with the sensitive and the intellectual is reducible in the main to this all-important rule: *carefully to prevent the principles proper to sensitive apprehension from passing their boundaries and meddling with the intellectual.* In every judgment, as intellectually enunciated, the predicate is a condition without which the subject is asserted to be unthinkable. The predicate is, in short, the principle in terms of which the subject is known. If, then, it be a concept of sense, it will be a condition only of possible sensitive apprehension, and will be in full accord only with a subject of judgment whose concept is likewise sensitive. If it be applied to an intellectual concept, the judgment so constituted will be valid only according to subjective laws. It will not be predicable, nor capable of being objectively stated, of the intellectual notion itself. It can be asserted *only as a condition without which there is no possibility of sensitive knowledge of the given concept.*[1] Since the illusions of the intellect in decking out sensitive concepts as intellectual marks may be called a fallacy of subreption (by analogy with the accepted meaning of the term subreption), the interchange of the intellectual and the sensitive will be the metaphysical fallacy of subreption (the fallacy of intellectualising the phenomenon, if a barbarism may be forgiven). Such hybrid axioms (hybrid, in that they proffer what is sensitive as being necessarily bound up with the intellectual concept), I call a surreptitious axiom. Those principles of intellectual error which have most harmfully infested metaphysics have, indeed, proceeded from these spurious axioms.

But we must go more deeply into this question, in order that we may have something ready to hand and clearly knowable, to serve as a criterion and touchstone of these judgments, distinguishing them from those that are genuine. And at the same time, since these judgments may

[1] The use of this criterion is easy and fruitful in distinguishing principles which enunciate only laws of sensitive knowledge from those which propound something further about the objects themselves. For if the predicate be an intellectual concept, the reference of it to the subject of the judgment (however sensitively that subject may be apprehended) indicates always a mark belonging to the object itself. But if the predicate be a sensitive concept, then, since the laws of sensitive apprehension are not conditions of the possibility of things themselves, it will not hold of the subject of the judgment as intellectually conceived, and so will not be capable of being enunciated objectively. Thus, since in that common axiom "whatever exists is somewhere," the predicate contains conditions of sensitive apprehension, it will not be capable of being universally enunciated of the subject of the judgment, viz., of every *existent;* and so this formula, taken as having objective reference, is false. But, if the proposition be converted so that an intellectual concept becomes predicate, a completely true proposition will emerge, namely: "Whatever is somewhere, exists."

perhaps seem to be firmly rooted in the intellect, we also require an art of assaying by the aid of which a just estimation may be made of how much belongs to the sensitive and how much to the intellectual.

§ 25. The principle, then, of the reduction of any surreptitious axiom is this: *if there be universally predicated of any intellectual concept anything belonging to the relations of space and time, it must not be enunciated as objective; it denotes only the condition without which the given concept is not knowable by sense.* That an axiom of the surreptitious type is spurious and, if not false, at least rashly and precariously asserted is clear from the fact that the subject of the judgment, being intellectually conceived, belongs to the object, whereas the predicate, as containing determinations of space and time belongs only to the conditions of human sensitive knowledge. Since this latter knowledge does not necessarily qualify every cognition of the object, it cannot be universally enunciated of the given intellectual concept. The reason why the intellect is so liable to this fallacy of subreption is that the deception takes place under cover of another rule which is genuine enough. We rightly suppose that what cannot be known through any intuition whatever is not thinkable at all, and so is impossible. But since by no effort of the mind can we attain even in imagination another kind of intuition than that which takes place according to the form of space and time, it comes about that we regard as impossible all intuition whatever which is not bound by these laws (neglecting the pure intellectual intuition which is exempt from the laws of the senses, such as the divine intuition which Plato calls an idea), and so subject all possible things to the sensitive axioms of space and time.

§ 26. All the illusions of sensitive knowledge masquerading as intellectual knowledge—the illusions from which arise the surreptitious axioms—can be reduced to three kinds, of which the following are the general formulae:

1. The sensitive condition, under which alone the intuition of an object is possible, is a condition of the possibility of the object itself.

2. The sensitive condition, under which alone data can be compared for the formation of an intellectual concept of an object, is a condition of the possibility of the object itself.

3. The sensitive condition, through which alone the subsumption of any presented object under a given intellectual concept is possible, is also a condition of the possibility of the object itself.

§ 27. A surreptitious axiom of the first class is: *Whatever is, is somewhere and somewhen.*[1] By this spurious principle all beings, even if intellectually apprehended, are bound in their existence by conditions of space and time. Thence are tossed to and fro those idle questions as to the position in the corporeal universe of immaterial substances (of which, however, as immaterial, no sensitive in-

[1] Space and time are conceived as comprehending in themselves all things which can in any way meet the senses. Thus, according to the laws of the human mind, no intuition of any being is given except as that being is contained in space and time. With this prejudice there can be compared another, not properly a surreptitious axiom, but a trick of the imagination, which might be expressed in the general formula: in whatever exists there is space and time, i.e., all substance is extended and in continual change. For though all those whose conceptions are somewhat gross are firmly tied down by this law of the imagination, they themselves easily see that this applies only to the efforts of the imagination in its attempts to depict to itself the forms of things, not to the conditions of existence.

tuition is given, nor any representation under such [spatial] form), as to the seat of the soul, and other questions of the kind. The sensible and the intellectual being thus improperly confounded, like square and round, it commonly happens that one of the disputants appears as it were to be milking a he-goat and the other to be holding a sieve. The presence of immaterial things in the corporeal world is, however, virtual, not local, though it may improperly be so called. Space contains the conditions of possible reciprocal actions only between material bodies. What constitutes the external relations of force [or influence] between one immaterial substance and another or between immaterial substances and bodies is a problem quite beyond human intellect, as the truly perspicacious Euler, who was for the rest a great investigator and judge of phenomena, acutely noted in his letters to a German princess.[1]

When philosophers come to the concept of a supreme and extramundane being, words cannot express the extent to which they are befooled by these shadows that flit before the intellect. They imagine the presence of God to be local, and involve God in the world as if He were comprised in infinite space all at once, although they themselves have to make up for this limitation by conceiving the locality as it were *per eminentiam*, namely, as infinite. But to be in a number of places at the same time is absolutely impossible, because different places are outside one another, and consequently what is in more than one place is outside itself and externally present to itself, which is self-contradictory.

[1] [*Lettres à une princesse d'Allemagne sur divers sujets de Physique et de Philosophie* (3 vols., 1768–72); see letters 92, 93.]

As regards time, philosophers involve themselves in an inextricable maze, for not only do they disconnect it from the laws of sensitive apprehension, they likewise carry it beyond the confines of the world, to the extramundane being itself, as a condition of the existence of such being. Hence the absurd questions with which they torment their minds, e.g., why God did not fashion the world many ages earlier. They persuade themselves that it can easily be conceived how God perceives things present, that is, the actual things of the time in which He is; but how He foresees the future, that is, the actual things of a time in which He is not yet, they think difficult of understanding. As if the existence of a necessary being must descend through all the moments of imaginary time, and as though, part of His duration already exhausted, He looked forward to the eternity He is yet to pass through and to the events of the world to take place during that time! All these difficulties vanish like smoke when the notion of time is rightly understood.

§ 28. Preconceptions of the second kind, since they delude the intellect through the sensitive conditions by which the mind is bound if it wishes to arrive at an intellectual concept in certain cases, are still more deeply hidden. One of these generally affects the knowledge of quantity, the other that of qualities. The former is: *every actual aggregate can be expressed by a number, and therefore every quantum is finite.* The latter is: *whatever is impossible is self-contradictory.* In neither case does the concept of time enter into the predicate notion itself, nor is it considered a mark of the subject, but it serves as the medium in which the concept of the predicate is formed, and so affects the intellectual concept of the subject as a condition, inasmuch as we attain the concept only by its aid.

As regards the former axiom, since every quantum and every series is distinctly known only by successive co-ordination, the intellectual concept of a quantity and an aggregate arises only by the help of this concept of time, and never attains completeness save where the synthesis can be completed in a finite time. Owing, therefore, to the limits of our intellect, an infinite series of co-ordinates cannot be distinctly comprehended, and accordingly, through a fallacy of subreption, appears impossible. Certainly, according to the laws of pure intellect every series of effects has its assignable *ground of existence*[1] (i.e., in a series of effects regress without limit is not possible), while according to the laws of sensibility every series of co-ordinates has an assignable *beginning*. But these propositions, the latter involving the measurability of the series, the former the dependence of the whole, are wrongly held to be identical. In similar fashion, when the intellect proves that given a complex of substances there are given also elements of composition, i.e., simples, there is combined with the argument a false principle, borrowed from sensitive knowledge, namely, that in such a compound there cannot be an infinite regress in the composition of parts, i.e., that in every compound there is a definite number of parts, a proposition whose meaning is certainly not on all fours with that of the former, and which it is therefore rash to substitute for it. That the world, therefore, is limited in its quantity (and not a maximum), that it implies a first principle [of its being], that bodies consist of simples, can indeed be

[1] [*principium*.]

known on the trustworthy authority of reason. But that the universe with respect to its mass is mathematically finite, that its age is assignable in terms of a unit measure, that there is a definite number of simples constituting any body, are propositions which openly declare their origin from the nature of sensitive knowledge; and whatever truth may otherwise belong to them, they nevertheless labour under the indubitable defects of their origin.

As regards the second surreptitious axiom, it arises from rashly converting the principle of contradiction. To that primary principle the concept of time adheres to this extent, that, given contradictorily opposed characters in the same thing *at the same time*, an impossibility is manifest, and is thus enunciated: *whatever is and is not at the same time is impossible.* Since something is here predicated through the intellect in a case which is given according to the laws of sensibility, the judgment is perfectly true and evident. But if the axiom be converted and it be asserted: *everything impossible both is and is not at the same time* (i.e., involves a contradiction), something is predicated universally of an object of reason, through sensitive knowledge, and the intellectual concept of the possible and impossible is thereby subjected to conditions of sensitive knowledge, namely to relations of time. This is indeed altogether true of the laws by which the human mind is bound and limited; but when the statement is made objectively and universally, it can in no wise be granted. That is to say, our intellect recognizes impossibility only where it can remark the simultaneous enunciation of opposites about the same subject, i.e., only where a contradiction occurs. The human intellect, therefore, can make no judgment of impossibility in cases in which this contradiction is not found. But it is a hasty conclusion, and involves the regarding of subjective conditions of judgment as objective, to say that no intellect whatsoever can do so, and that what does not involve a contradiction is, for that reason, possible. To this source is due the host of fictitious forces fabricated at will, which, not finding any obstacle in the principle of contradiction, are poured forth in multitudes by those of speculative mind, or, if the description be preferred, by those whose minds are given to chimeras. For since force is nothing but the relation of a substance A to something else B (an accident) as ground to consequent, the possibility of any force *does not rest upon the identity* of cause and effect or of substance and accident, and consequently the impossibility of falsely imagined forces *does not depend exclusively on contradiction.* It is, therefore, illegitimate to assume any primary force as possible, unless it be given by experience; by no acuteness of intellect can its possibility be apprehended *a priori.*

§ 29. The third species of surreptitious axioms, while arising from conditions special to the subject, and thence rashly carried over to objects, do not (as the second kind do) originate in our making the sensitively given [i.e., time and space] the only path to the intellectual concept; but in our holding that only by their aid can it be applied to a case given in experience. That is to say, we maintain that only through them can it be known whether or not anything is contained under a certain intellectual concept. Of this type is the trite maxim of certain schools: *whatever exists contingently has at some time not existed.* This spurious principle arises from the poverty of the intellect, which for the most part discerns the nominal marks of contingency or necessity, seldom the real ones. Since, therefore, we can scarcely hope to determine, through marks deduced *a priori*, whether the opposite of some substance is possible, we shall be able to do so only in so far as we have evidence that at one time the substance did not exist. It is truer that changes prove contingency than that contingency proves mutability; and accordingly, if nothing transient and perishable met us in the world, we could hardly have acquired any notion of contingency. Thus, while the direct proposition, *whatever at some time was not, is contingent*, is true, its converse only points to conditions under which alone we can discern whether a thing exists necessarily or contingently. If, therefore, it be enunciated as a subjective law (as it really is), it should be expressed thus: *if there be no evidence that there was a time when a certain thing did not exist, the common intelligence does not supply sufficient marks of its contingency*. This subjective law is later tacitly converted into an objective condition, as if without this condition there were no room at all for contingency. This being done, a spurious and erroneous axiom arises. For though this world exists contingently, it is everlasting, i.e., simultaneous with every moment of time; and it would, therefore, be erroneous to assert that there ever was a time when it did not exist.

§ 30. With the surreptitious principles there are closely associated certain others, which do not indeed communicate to the given intellectual concept any stain of sensitive apprehension, but which yet so play upon the intellect that it regards them as arguments drawn from the object. They are really commended to us only by the special nature of the intellect, owing to their convenience for its free and extended employment; and so, equally with those already enumerated, they rest upon subjective grounds, though, in this case, upon laws of intellectual, not of sensitive, knowledge; namely, upon the conditions under which the intellect seems to itself to make easy and ready use of its insight.

Let me here, by way of conclusion, add some remarks upon these principles. So far as I know, they have never yet been clearly expounded. I call principles of convenience those rules of judgment, to which we willingly submit and to which we cling as if they were axioms, solely for the reason that, *if we gave them up, scarcely any judgment about a given object would be possible to our intellect.* Into this class fall the following principles: (1) that *all things in the universe happen according to the order of nature*—a principle which Epicurus affirmed without restriction, and which all philosophers with one voice assert as holding good, subject only to those very rare exceptions which may have to be admitted under stress of the most extreme necessity. We do not assume it because of our possessing so wide a knowledge of cosmic happenings according to the common laws of nature, or because we perceive the impossibility, or the minimal hypothetical possibility, of the supernatural. We do so because, if we departed from the order of nature, there would be no use at all for intellect, and because the hasty appeal to the supernatural is a couch upon which the intellect slothfully reclines. For the same reason, in our exposition of phenomena, we are careful to exclude *comparative miracles*.[1] I mean the

[1] [*miracula comparativa.*]

influence of spirits, since, as their nature is unknown to us, it would be greatly to the detriment of the intellect if it were diverted from the light of experience—through which alone it has the means of judging between rival interpretations—to the shadows of forms and causes to us unknown.

(2) The second principle is due to that preference for unity, so characteristic of the philosophic mind. From it has issued that commonly accepted canon: *principles are not to be multiplied beyond what is absolutely necessary.* We accept this principle not because we have insight into causal unity in the world, either by reason or by experience; we are driven to search for it by the impulsion of the intellect, which believes itself to have succeeded in the explanation of phenomena only in so far as it is allowed to descend from one and the same principle to the greatest number of consequences.

(3) The third principle of this kind is that *nothing material either comes into being or perishes*, and that all the changes of the world concern only its form. This postulate, on the recommendation of the common intellect, has pervaded all Schools of philosophy, not that it has been regarded as an ascertained fact, or as demonstrated by arguments a priori; but because, if it were admitted that matter itself is transient and perishable, nothing stable and lasting would be left to serve in the further explanation of phenomena according to universal and constant laws, and so in the further use of the intellect.

So much then as regards method, especially in reference to the distinction of sensitive and intellectual knowledge. Should this method, by a more thorough investigation, ever be reduced to rule, it will occupy the place of a propaedeutic science, to the immense benefit of all who would explore the innermost recesses of metaphysics.

NOTE.—Since in this last section the investigation of method occupies all the space available, and the rules stating the true form of reasoning in regard to the sensitive are evident of themselves, and do not borrow their evidence from the examples introduced by way of illustration, I have spoken of these examples only, as it were, in passing. So it is not strange if it will seem to many that much has there been asserted with more boldness than truth, much indeed that will demand a greater weight of proof when an opportunity offers of more extended treatment. Thus what I have said in § 27 regarding the locality of immaterial things needs explanation, which the reader may, if he please, seek in Euler, *loc. cit.*, tome. 2, pp. 49–52. For it is not the case that the soul is in interaction with the body because it is confined in a certain part of it, but a definite location in the universe is attributed to it because it is in reciprocal interaction with a certain body and loses all location in space when this interaction ceases. Thus its locality is derivative, and belongs to it only contingently; it is not a primal condition that of necessity goes along with its existence. For whatever things cannot in themselves be objects of the outer senses as found in man, that is to say, immaterial things, are altogether exempt from the universal condition of objects of outer sense, namely, space. Accordingly, while absolute and immediate locality can be denied of the soul, a hypothetical and mediate locality can be attributed to it.